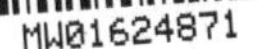

INTRODUCING

MATH! GRADE 4

ARGOPREP

TEACHER RECOMMENDED

TOPICS COVERED

PRACTICE MAKES PERFECT

- Operations & Algebraic Thinking
- Numbers and Operation in Base Ten
- Fractions
- Measurement and Data
- Geometry

ArgoPrep is one of the leading providers of supplemental educational products and services. We offer affordable and effective test prep solutions to educators, parents and students. Learning should be fun and easy! For that reason, most of our workbooks come with detailed video answer explanations taught by one of our fabulous instructors.

Our goal is to make your life easier, so let us know how we can help you by e-mailing us at: info@argoprep.com.

ISBN: 9781946755773
Published by Argo Brothers, Inc.

Aknowlegments:
Icons made by Freepik, Creaticca Creative Agency, Pixel perfect , Pixel Buddha, Smashicons, Twitter , Good Ware, Smalllikeart, Nikita Golubev, monkik, DinosoftLabs, Icon Pond from www.flaticon.com

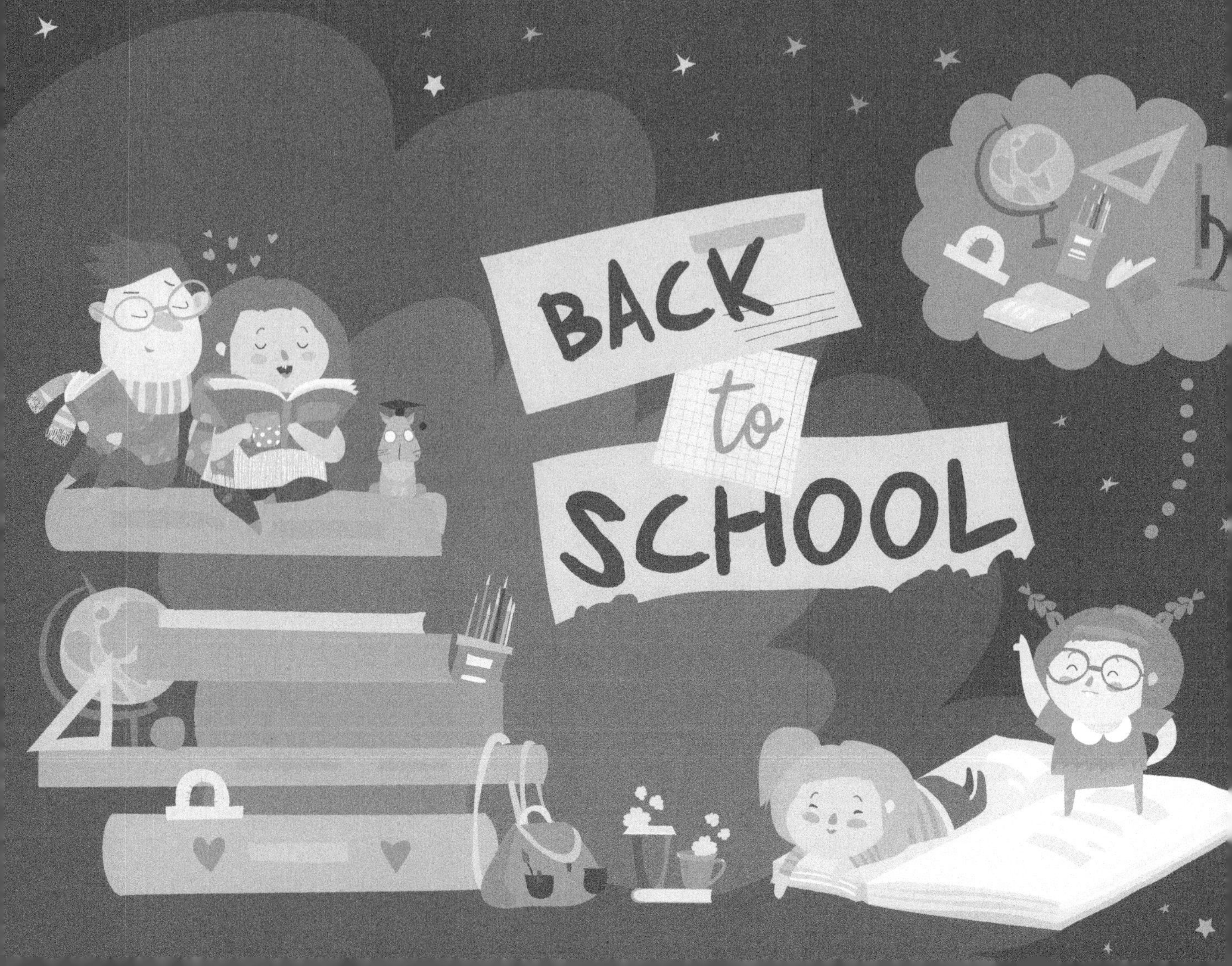

BACK to SCHOOL

- ArgoPrep is a recipient of the prestigious **Mom's Choice Award**.

- ArgoPrep also received the 2019 **Seal of Approval** from Homeschool.com for our award-winning workbooks.

- ArgoPrep was awarded the 2019 **National Parenting Products Award, Gold Medal Parent's Choice Award** and **the Tillywig Brain Child Award.**

Want an amazing offer from ArgoPrep?

7 DAY ACCESS

to our online premium content at **www.argoprep.com**

Online premium content includes practice quizzes and drills with video explanations and an automatic grading system.

Chat with us live at **www.argoprep.com** for this exclusive offer.

TABLE OF CONTENTS

HOW TO USE THE BOOK

Welcome to the **Introducing Math!** series by ArgoPrep.
This workbook is designed to provide you with a comprehensive overview of Grade 4 mathematics.

While working through this workbook, be sure to read the topic overview that will give you a general foundation of the concept. At the end of each chapter, there is a chapter test that will assess how well you understood the topics presented.

This workbook comes with free digital video explanations that you can access on our website. If you are unsure on how to answer a question, we strongly recommend watching the video explanations as they will reinforce the fundamental concepts.

We strive to provide you with an amazing learning experience. If you have any suggestions or need further assistance, don't hesitate to email us at info@argoprep.com or chat with us live on our website at www.argoprep.com

HOW TO WATCH VIDEO EXPLANATIONS

IT IS ABSOLUTELY FREE

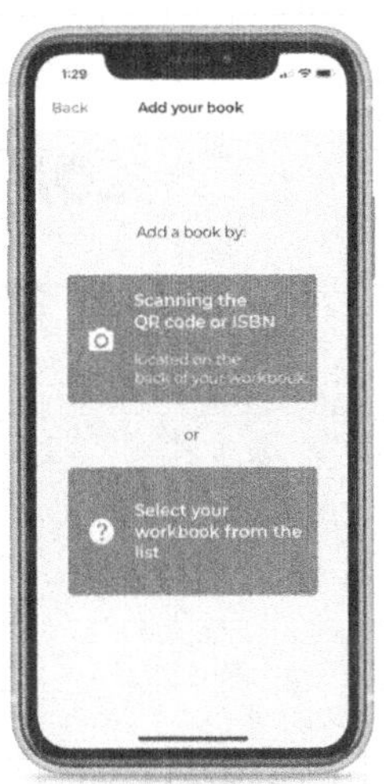

Download our app:
ArgoPrep Video Explanations
to access videos on any mobile device or tablet.

or

Step 1 - Visit our website at: www.argoprep.com/k8
Step 2 - Click on the Video Explanations button located on the top right corner.
Step 3 - Choose the workbook you have and enjoy video explanations.

OTHER BOOKS BY ARGOPREP

Here are some other test prep workbooks by ArgoPrep you may be interested in. All of our workbooks come equipped with detailed video explanations to make your learning experience a breeze! Visit us at **www.argoprep.com**

COMMON CORE MATH SERIES

COMMON CORE ELA SERIES

INTRODUCING MATH!

Introducing Math! by ArgoPrep is an award-winning series created by certified teachers to provide students with high-quality practice problems. Our workbooks include topic overviews with instruction, practice questions, answer explanations along with digital access to video explanations. Practice in confidence - with ArgoPrep!

SCIENCE SERIES

Science Daily Practice Workbook by ArgoPrep is an award-winning series created by certified science teachers to help build mastery of foundational science skills. Our workbooks explore science topics in depth with ArgoPrep s 5 E S to build science mastery.

KIDS SUMMER ACADEMY SERIES

ArgoPrep's **Kids Summer Academy** series helps prevent summer learning loss and gets students ready for their new school year by reinforcing core foundations in math, english and science. Our workbooks also introduce new concepts so students can get a head start and be on top of their game for the new school year!

SUMMER ACTIVITY PLAYGROUND SERIES

Summer Activity Playground is another summer series that is designed to prevent summer learning loss and prepares students for the new school year. Students will be able to practice math, ELA, science, social studies and more! This is a new released series that offers the latest aligned learning standards for each grade.

FrogYogi Math Series

It combines fun and engaging activities along with math concepts. Your child will be hooked in learning math on a daily basis. This workbook includes fun and effective yoga math breaks between solving problems helping the brain relax and retain more information.

Chapter 1: Operations & Algebraic Thinking

You can solve multiplication problems in a variety of ways. One way to consider multiplication problems is to break them down into comparison problems.

Consider the problem **2 x 4 = 8**. This problem can be expressed as **4** groups of **2**:

2 + 2 + 2 + 2

Or as **2** groups of **4**:

4 + 4

Both of which equal **8**.

Another way to consider this problem is by comparing the two factors in terms of the answer:

The number 8 is four times as large as 2.

What does that mean exactly?

If I take the number 2, and make it larger by 2 four times, then I end up with the product of 8.

Let's look at another one:

5 x 2 = 10

How can I break that into groups?

5 + 5 OR **2 + 2 + 2 + 2 + 2**

What if I compare the two factors?

The number 10 is 5 times as large as the number 2.

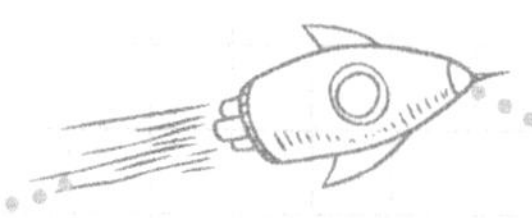
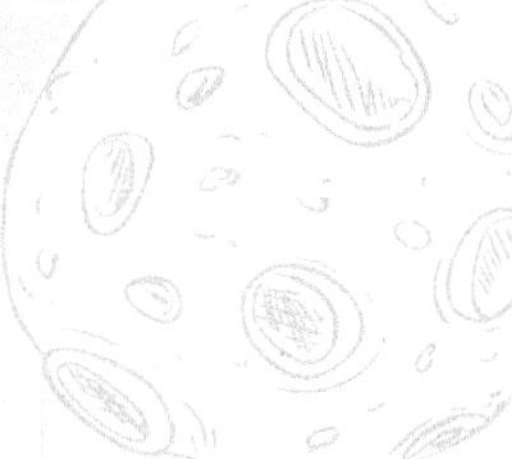

1. The number ____________ is 6 times as large as 4.

A. 24

B. 30

C. 28

D. 32

SHOW YOUR WORK

2. The number 56 is ____________ times as large as 7.

3. 45 is 9 times as large as what number?

A. 10

B. 8

C. 6

D. 5

SHOW YOUR WORK

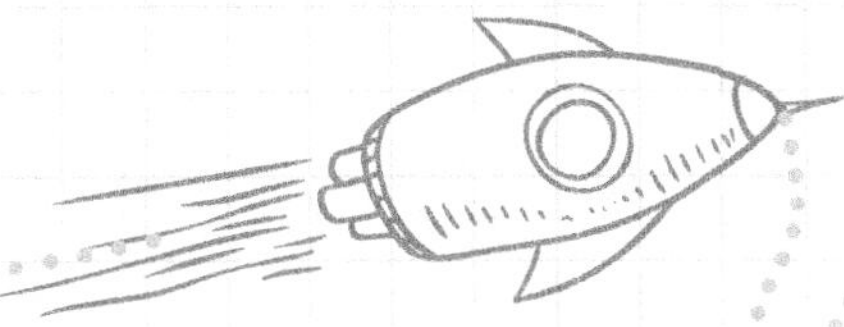

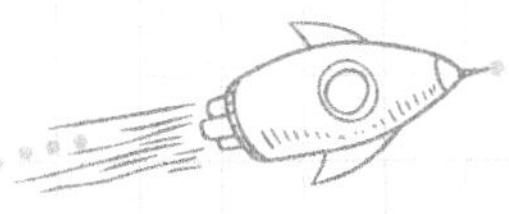

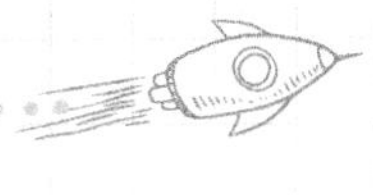

4. The number 21 is ____________ times as large as ____________.

SHOW YOUR WORK

5. 36 is 9 times as large as what number?

A. 3

B. 4

C. 5

D. 6

SHOW YOUR WORK

6. Evan and Morgan sell candy bars for their band. Evan sells 12 candy bars. Morgan sells four times as many candy bars as Evan. What problem represents how many candy bars Morgan sells?

A. 12 x ?= 4

B. 4 x ? = 12

C. 12 x 4 = ?

D. 12 + 4 =?

SHOW YOUR WORK

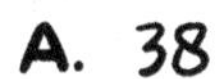

7. Using the information in the problem above, if Jack sold 10 more candy bars than Morgan, how many candy bars did Jack sell?

A. 38

B. 45

C. 48

D. 58

SHOW YOUR WORK

8. Alicia and Mary collect stamps. Alicia has 15 stamps. Mary has twice as many stamps as Alicia. What problem represents how many stamps Mary and Alicia have?

A. 15 x 2 = ?

B. (15 x 2) - 15 = ?

C. (15 x 2) + 15 = ?

D. 15 + 15 + 2 = ?

SHOW YOUR WORK

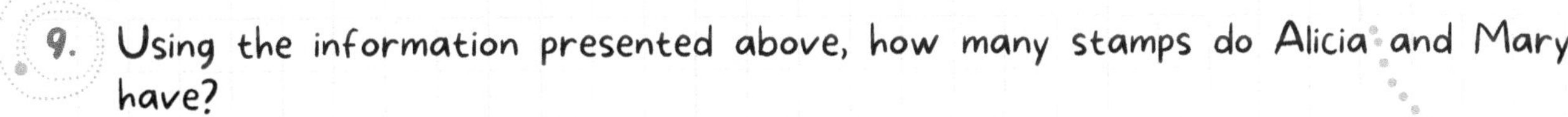

9. Using the information presented above, how many stamps do Alicia and Mary have?

A. 20 stamps

B. 25 stamps

C. 30 stamps

D. 45 stamps

SHOW YOUR WORK

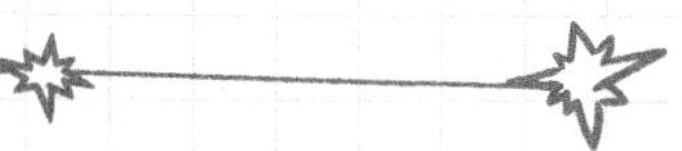

10. Juan and Marco receive an allowance every week. Juan receives three times as much money as Marco. Marco receives **$2**. How much money does Juan receive? In a few sentences, explain how you calculated how much money Juan receives.

SHOW YOUR WORK

NOTES

You can apply the comparison skills we have worked on when you solve word problems involving multiplication and division. **Remember, when you have a word problem, you should break the problem first into what you know and then into what you need to know.** We can use comparisons to help find what we need to know.

Let's look at an example.

Chris got four toys for his birthday. Each toy needs 2 batteries to operate. How many batteries does he need in all?

Let's dig into the problem. First, what do we know?

We know Chris has 4 toys. We know each toy needs two batteries.
What do we want to know?
We want to know how many batteries he needs in all?
We can draw a picture to help us illustrate how many batteries we need.

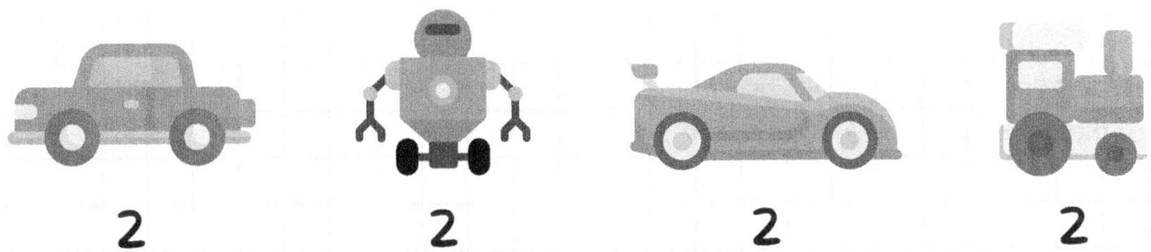

2 2 2 2

For all the toys to work, he needs 8 batteries!
We can also solve this problem by comparison. What number is 2 times as large as 4? 8

Let's look at another one!

Michelle orders 3 pizzas. Each pizza can give her 8 slices. How many slices of pizza does she have in all?

Let's look at a picture.

8 8 8

Michelle has 24 slices of pizza! We can also solve this problem by comparison. What number is 8 times as large as 3? 24

It works for division too!

Our classroom has 36 books in its library. There are four groups of students. How many books should each group take from the library?

If I circle four equal groups from the set of books above, I end up with 9 books in each group.

We can also solve this problem by comparison. Thirty six is 4 times as large as what number? 9

1.1.B Use the four operations with whole numbers to solve problems

1. Jacob rides his bike to school every day for 7 minutes. On the weekend, he rides his bike for three times as long. What problem represents how long he rides his bike on weekends?

A. $7 \times 3 = ?$

B. $? \times 3 = 7$

C. $3 \times ? = 7$

D. $21 \times 3 = ?$

2. Elizabeth eats three pieces of fruit a day. Each member of her family also eats three pieces of fruit. If her family consumes 21 pieces of fruit a day, what problem represents how many people are in her family?

A. $21 \times 3 = 7$

B. $3 \times ? = 21$

C. $7 \times ? = 21$

D. $3 \times 7 = 21$

3. Emma collects stickers. She buys a new pack of stickers from the store. Each sheet of stickers has 12 stickers and the pack contains 6 total sheets. What problem represents how many stickers she adds to her collection?

A. $6 \times ? = 12$

B. $12 \times ? = 6$

C. $12 \times 6 = ?$

D. $6 \times 2 = ?$

1.1.B | Use the four operations with whole numbers to solve problems

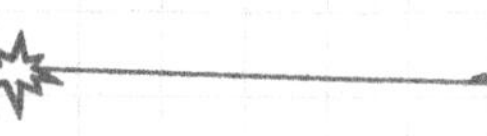

4. Surjo does additional math practice every day. He does **63** total problems a week. Write a problem that represents the number of problems he completes daily in a 7 day week.

SHOW YOUR WORK

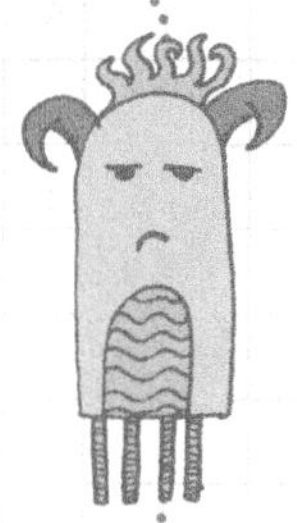

5. Parker raised **$96** for the food bank which was twice as much as his classmate Liam raised. What problem represents how much money Liam raised?

A. 96 x ? = 2

B. 96 x 2= ?

C. 48 x 2 = ?

D. 2 x ? = 96

SHOW YOUR WORK

6. Thomas purchased nine shirts for a total of **$63**. How much did each shirt cost if all shirts were priced the same?

A. $7

B. $6

C. $7

D. $3

SHOW YOUR WORK

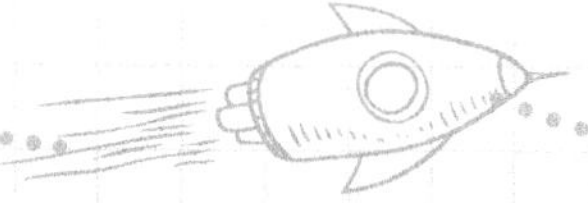

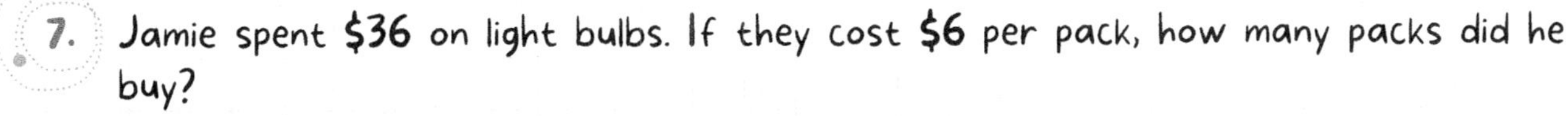

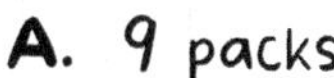

7. Jamie spent $36 on light bulbs. If they cost $6 per pack, how many packs did he buy?

A. 9 packs

B. 6 packs

C. 7 packs

D. 8 packs

SHOW YOUR WORK

8. Damon bought 320 cookies. He had put them into 8 boxes evenly. How many cookies did Damon put in each box?

A. 44 cookies

B. 30 cookies

C. 40 cookies

D. 36 cookies

SHOW YOUR WORK

9. Grace scored 96 points, which was 3 times more than Melissa. How many points did Melissa score?

A. 31

B. 32

C. 33

D. 34

SHOW YOUR WORK

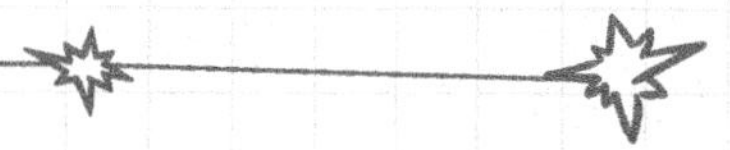

10. Jane has 8 candies and 9 cupcakes in each box. If she has 7 boxes, how many candies and cupcakes does she have altogether?

SHOW YOUR WORK

NOTES

We can apply the skills used above when we solve word problems involving more than one step. First, figure out what we know and then figure out what we want to know. What problem do we have to solve first to figure out what we want to know?

Let's look at an example.

Michelle orders 3 pizzas. Each pizza can give her 8 slices. If she is feeding 6 people, how many slices of pizza can each person have?
Here is what we know: 3 pizzas, 8 slices in each pizza to feed 6 people.

We want to know the number of slices per person so first we need to figure out how many slices total. Remember, what number is 8 times as large as 3? Our answer is 24, so we know has 24 slices to split 6 ways.

Twenty four is 6 times as large as what number? 4 so each person can have 4 pieces of pizza.

We can also use a letter to stand for an unknown quantity.

Let's look at the problem above.

What number is 8 times as large as 3? Another way to say this is $8 \times 3 = n$
Twenty four is 6 times as large as what number?
Another way to say this is $24 = 6 \times n$

In both those problems, the n takes the place of the number we don't know.

What if we get an answer with a remainder?

We have to determine what answer is the most reasonable. In other words, what answer makes the most sentence based on what the problem is looking for.

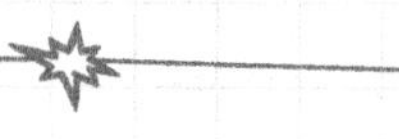

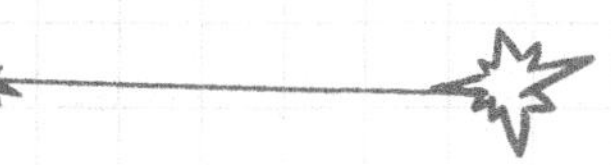

Let's look at an example.

Aliza's mom brings snacks to soccer practice. She wants to bring a sports drink. Currently, there is a sale for 6 bottles of sports drinks in a pack for $2.50. There are 16 people on Aliza's team. Assuming each person wants 1 bottle, how many packs of sports drink should she get?

To solve, first consider the problem. 6 times what number equals 16? You can also represent this as $6 \times n = 16$

When we consider this, we cannot find a whole number that works. If we use 2, we get 12 total bottles and when we use 3, we get 18 total bottles.

If each teammate wants 1 bottle, we need to get at least 16 bottles. So Aliza's mom should buy 3 packs so each teammate can get a bottle and then, she will have 2 bottles left over.

1. Our science club raised **$375** dollars from a fundraiser. They decided to each pay **$5** additional dollars to go on a field trip. The trip cost **$465** and was covered by the fundraiser and the additional pay. What problem do you need to solve first to determine how many students are in the club?

A. 465 - 375

B. 375 + 5

C. 465 + 375

D. 375 ÷ 5

SHOW YOUR WORK

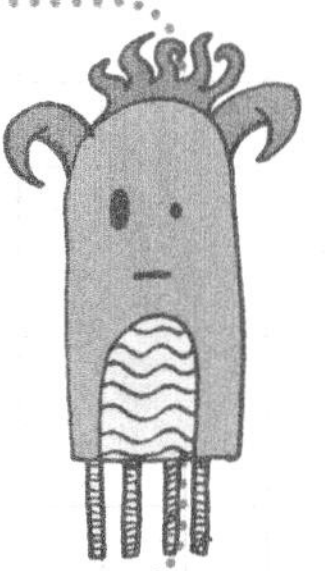

2. Using the information in problem 1, what do you do with the answer to determine how many students are in the club?

SHOW YOUR WORK

3. Freedom Elementary School is having a bake sale. Four parents have decided to donate **300** cupcakes, 1/3 of them chocolate. What problem needs to be solved first to determine how many chocolate cupcakes each parent needs to make?

A. 300 x ? = 100

B. 4 x ? = 300

C. 3 x 300 = ?

D. 3 x 100 =?

SHOW YOUR WORK

4. Jim and his seven friends decided to divide their restaurant bill evenly. If each person paid **$19**, then what was the total bill amount?

A. $2.71

B. $133

C. $134

D. $152

SHOW YOUR WORK

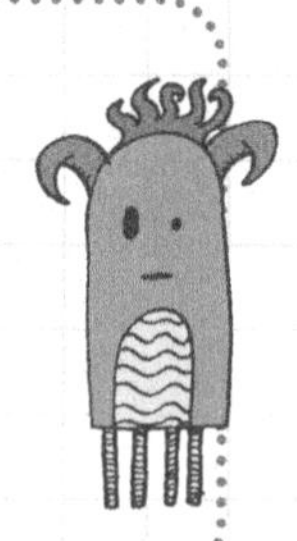

5. Mr. Lawrence needs markers for his class. He buys boxes of markers with **12** in each box. If he needs **87** markers, should he buy more boxes than he needs or fewer boxes than he needs? Why?

SHOW YOUR WORK

6. Julia spent **$5** on a burger for lunch. She then purchased two shirts that each cost **$17**. If she initially started with **$50**, how much money does she now have left?

A. $45

B. $28

C. $16

D. $11

SHOW YOUR WORK

7. Harry bought three bags of candy that had 75 pieces of candy in each bag. He plans to divide all the candy evenly among four friends and himself. How many pieces of candy will each person receive? Show your work.

SHOW YOUR WORK

8. Gabe earns $10 an hour delivering pizza and $11 an hour baby-sitting. Last week Gabe worked delivering pizzas for 7 hours and 3 hours for baby-sitting. How much money did he earn last week? Show your work.

SHOW YOUR WORK

9. Steve slept for 8 hours, which was 2 times as long as Tyler slept. Will slept 5 hours longer than Tyler. How many hours did Will sleep?

A. 4 hours

B. 5 hours

C. 8 hours

D. 9 hours

SHOW YOUR WORK

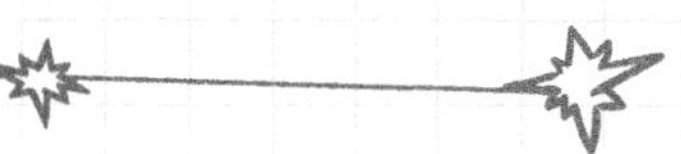

10. There were **15** necklaces that each had **32** beads. If Shana took all the beads off and put an equal amount on **7** chains, how many beads would be left over?

A. 0

B. 2

C. 4

D. 5

SHOW YOUR WORK

NOTES

1.2. Gain familiarity with factors and multiples

When reviewing the various operations, there are important terms to keep in mind. In multiplication, the numbers you multiply by each other are called **factors.** To get an answer (or product), you have to multiply at least two factors together. Most numbers can be broken down into at least two factors. When **multiplied** by other factors, you can generate multiples of a factor.

Let's look at an example:

The number 12 can be created by multiplying the two factors: 2 and 6. You can also say 12 is a multiple of 2 or a multiple of 6. You can list all the factors of a number like so:

12: 1, 2, 3, 4, 6, 12

You can get the number 12 by multiplying 1x12, 2x6 and 3x4.

Notice factors are whole numbers and are not decimal or fractions.

Something else to keep in mind. Factors can be prime numbers or **composite numbers.** If a factor can be broken down into a smaller set of whole numbers, we define it as a composite number. If a factor cannot be broken into a smaller set of whole numbers, we define it as **a prime number.**

Looking at our example of 12 above, the following factors are prime: 2 and 3. The following factors are composite: 4, 6 and 12.

1.2. Gain familiarity with factors and multiples

1. Which pair are factors of the number 64?

A. 8, 8

B. 8, 4

C. 2, 30

D. 4, 17

SHOW YOUR WORK

2. List the factors of 20. How do you find the factors of any given number?

SHOW YOUR WORK

3. Which factor of 18 is prime?

A. 1

B. 3

C. 9

D. 18

SHOW YOUR WORK

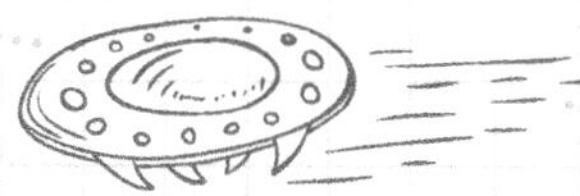

4. Which number is a multiple of 9?

A. 3

B. 4.5

C. 18

D. 20

SHOW YOUR WORK

5. Which product is the result of the factors 9 and 6?

A. 12

B. 36

C. 45

D. 54

SHOW YOUR WORK

6. Which pair are factors of the number 32?

A. 2, 15

B. 8, 4

C. 4, 4

D. 8, 5

SHOW YOUR WORK

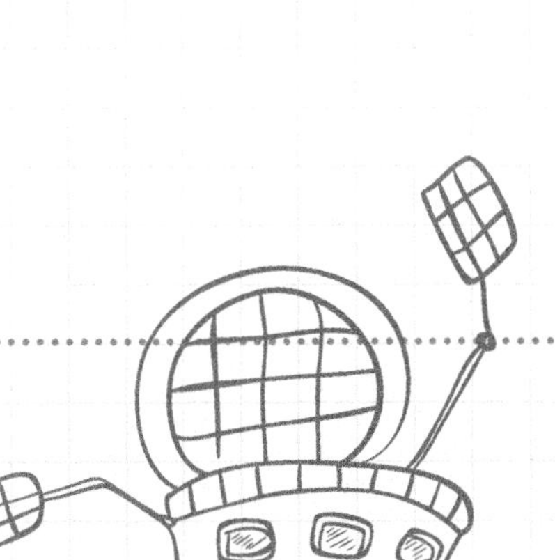

7. Which factor of 45 is prime?

A. 9

B. 15

C. 45

D. 5

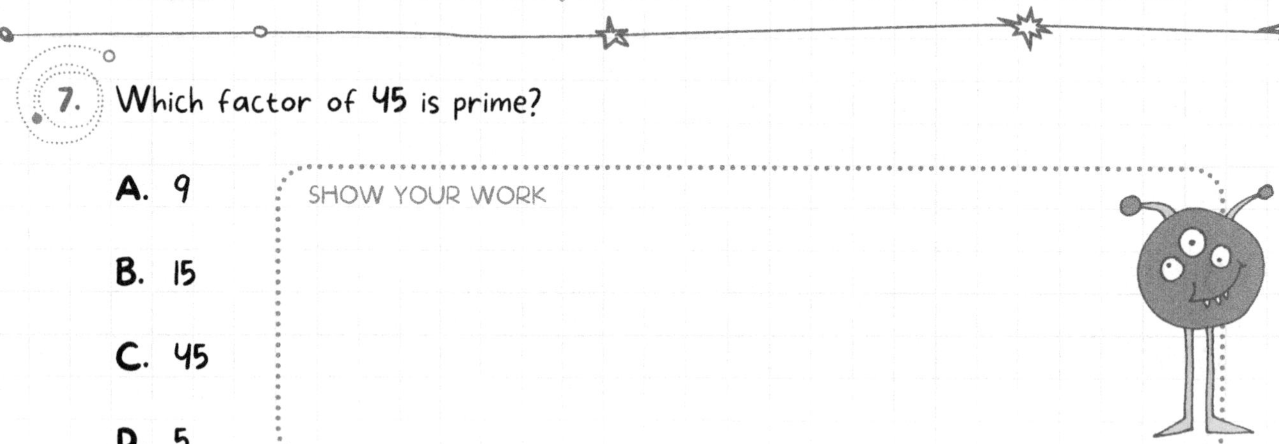

8. Why is 6 a composite number? Explain the difference between a composite number and prime number in your own words.

SHOW YOUR WORK

9. Which factor of 42 is composite?

A. 3

B. 7

C. 2

D. 21

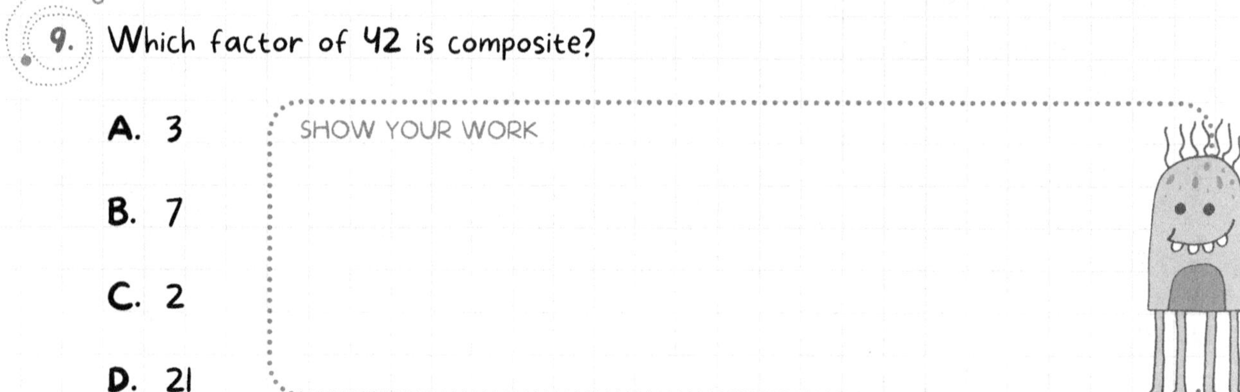

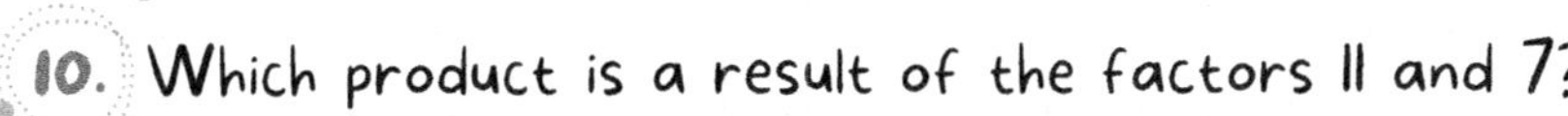

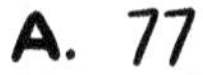

10. Which product is a result of the factors 11 and 7?

A. 77

B. 88

C. 99

D. 70

SHOW YOUR WORK

NOTES

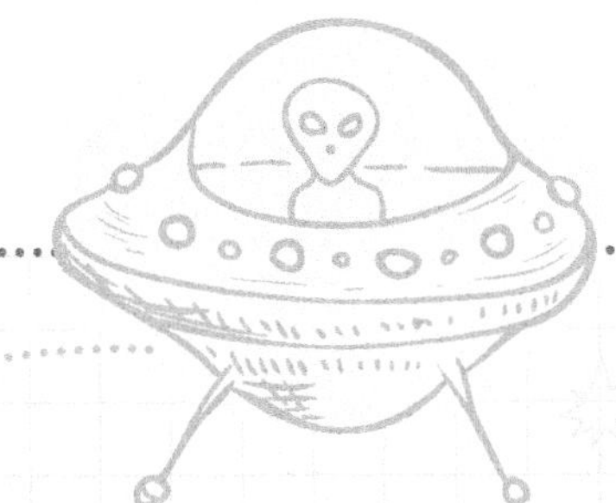

1.3. Generate and analyze patterns

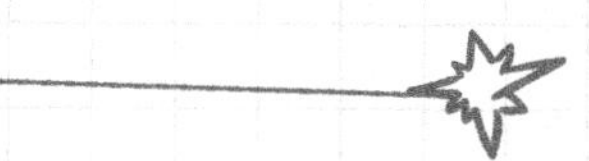

Have you noticed any patterns in what we have learned so far? Mathematics is full of patterns! Sometimes when you are solving problems, finding and applying the pattern can help you get farther. Let's review some information about patterns.

Here's an example pattern:

2, 4, 6, 8, 10, 12, 14

Can you find the **rule?** The rule of the pattern is what happens to each number. Often, patterns repeat a rule. The rule of the pattern above is + 2. Each **term** of the pattern is 2 more than the previous term. Following the rule, what would be the next term of the pattern? If you thought of 16, you got it!

1. What is the rule of the pattern below?

11, 9, 7, 5

A. + 2

B. - 2

C. + 3

D. - 3

SHOW YOUR WORK

2. What is the rule of the pattern below?

A. Triangle, Square

B. Square, Triangle

C. Square, Square, Triangle

D. Square, Rectangle

SHOW YOUR WORK

3. What is the rule of the pattern below?

10, 20, 40, 80, 160

A. + 10

B. + 40

C. x 2

D. - 20

SHOW YOUR WORK

4. Create a pattern that follows the rule of "minus 5."

SHOW YOUR WORK

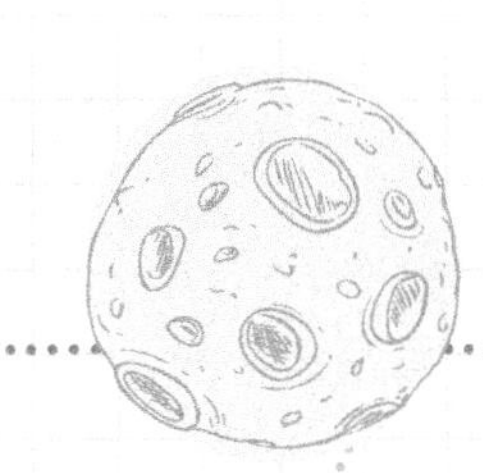

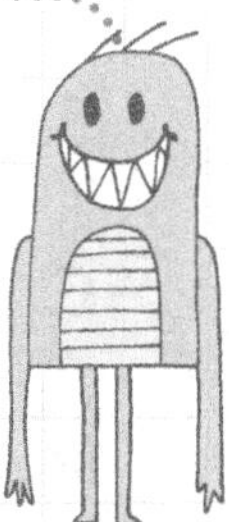

5. Create a pattern that follows the rule of "each additional shape adds a side."

SHOW YOUR WORK

6. What is the next term of the pattern below?

10, 7, 4,

A. 3

B. 2

C. 1

D. 0

SHOW YOUR WORK

7. What is the next term of the pattern below?

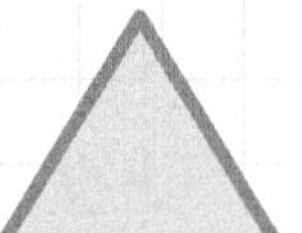  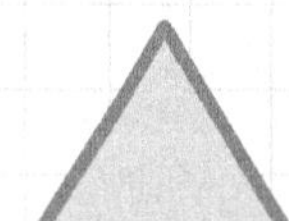  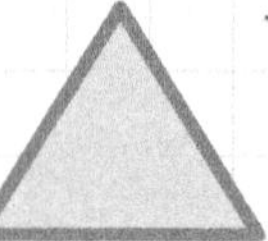

A. Triangle

B. Square

C. Circle

D. Diamond

SHOW YOUR WORK

8. What is the missing term in the pattern?

100, 90, 80, ? , 60

A. 70 B. 50 C. 40 D. 10

SHOW YOUR WORK

9. What is the missing term in the pattern?

3, 9, ?, 81, 243

A. 81 B. 27 C. 36 D. 729

SHOW YOUR WORK

10. What is the missing term in the pattern?

A. Plus sign B. Heart C. Lightning Rod D. Square

SHOW YOUR WORK

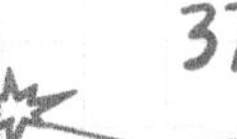

1.4. Chapter Test

1. What number is 7 times as large as 5?

A. 35

B. 42

C. 64

D. 12

SHOW YOUR WORK

2. What number is forty-eight, six times as large as?

A. 7

B. 8

C. 9

D. 10

SHOW YOUR WORK

3. Sara and Alicia collect flowers. On a walk, Sara collects 21 flowers. If Alicia collected three times as many flowers as Sarah, what problem represents how many flowers Alicia collected?

A. 21 x 3 = ?

B. 3 x ? = 21

C. 7 x ? = 21

D. 7 + 3 = ?

SHOW YOUR WORK

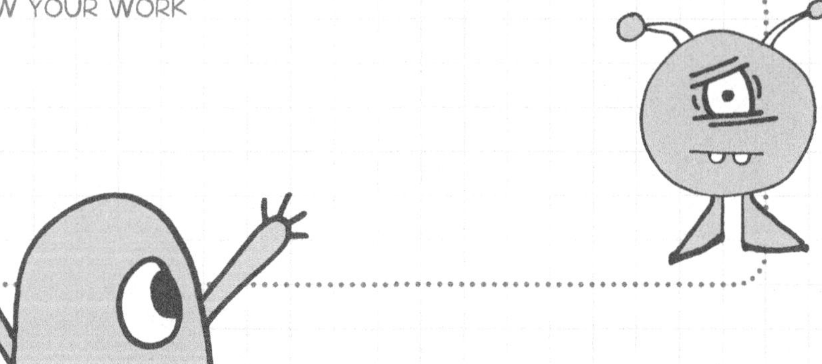

4. John eats the same breakfast most mornings. He usually has twice as much cereal as his little brother. If his little brother has a small 6 oz. bowl of cereal, how much cereal does John eat?

A. 6 oz **B.** 8 oz **C.** 12 oz **D.** 20 oz

SHOW YOUR WORK

5. Alice wants to buy a present for her mom's birthday. She has eight weeks of allowance. If she receives $3 every week for her allowance, which problem represents how much money she has to spend?

A. $8 + 3 = n$ **B.** $8 + n = 3$ **C.** $8 - 3 = n$ **D.** $8 \times 3 = n$

SHOW YOUR WORK

6. Mario wants to buy his dad a birthday present. He knows he wants to buy a new fishing rod for $47. He receives $5 from mowing his neighbor's yard every two weeks. How long will it take him to save money for his dad's present? Which problem do you need to solve first?

A. $n \times 5 = 47$ **B.** $47 \times n = 5$ **C.** $5 \times 2 = n$ **D.** $2 \times n = 5$

SHOW YOUR WORK

1.4. Chapter Test

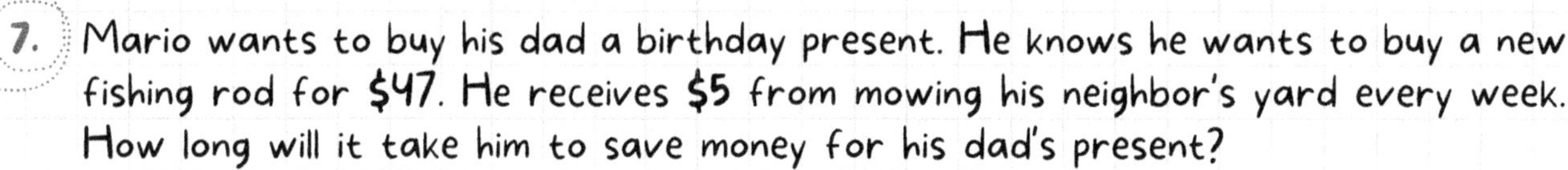

7. Mario wants to buy his dad a birthday present. He knows he wants to buy a new fishing rod for $47. He receives $5 from mowing his neighbor's yard every week. How long will it take him to save money for his dad's present?

A. 8 weeks

B. 9 weeks

C. 10 weeks

D. 11 weeks

8. Explain in a few sentences how you solved problem number 7.

SHOW YOUR WORK

9. Which two numbers are factors of 54?

A. 2, 16

B. 3, 16

C. 9, 7

D. 3, 18

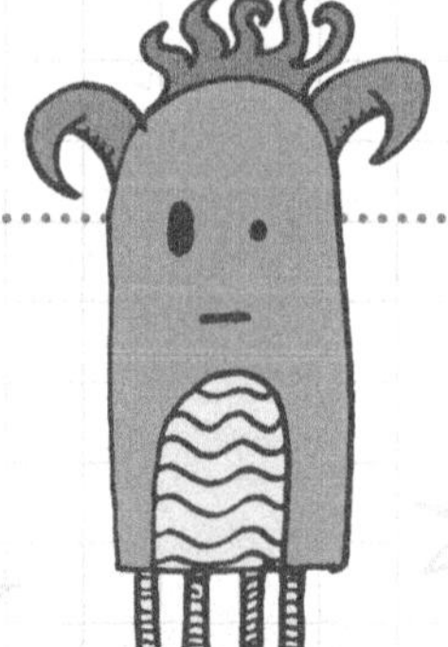

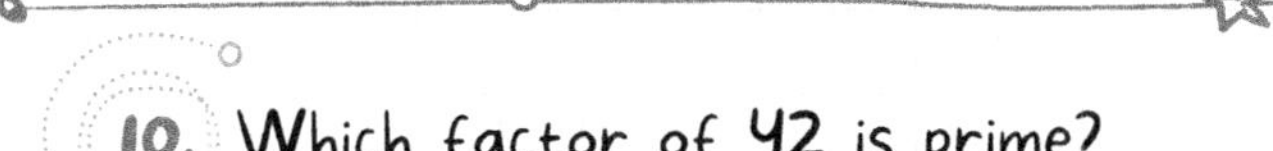

1.4. Chapter Test

10. Which factor of 42 is prime?

A. 6

B. 1

C. 14

D. 3

SHOW YOUR WORK

11. Which factor of 120 is composite?

A. 2

B. 3

C. 4

D. 5

SHOW YOUR WORK

12. Which answer shows multiples of 3?

A. 3, 9, 16

B. 6, 9, 12

C. 3, 1, 9

D. 6, 15, 26

SHOW YOUR WORK

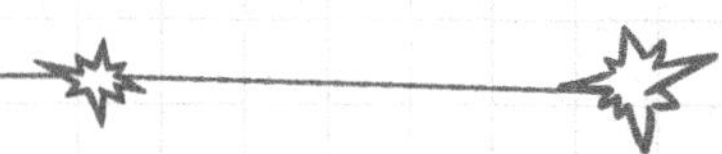

13. What is the pattern of the following set of numbers: 1, 10, 100, 1000

A. + 10

B. + 100

C. x 10

D. x 100

SHOW YOUR WORK

14. What is the next shape of the pattern below? How do you know?

SHOW YOUR WORK

15. What is the missing number of the following pattern?

4, 6, 12, 14, 28, ________, 60 ... How do you know?

SHOW YOUR WORK

NOTES

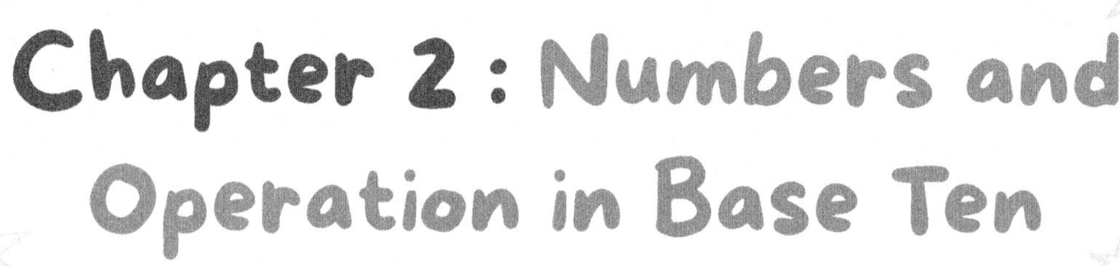

Chapter 2 : Numbers and Operation in Base Ten

Do you remember using base ten blocks when you were learning how to count? Lots of times, students use single counters and tens rods to represent numbers as they count from 1-20. When you are counting, numbers follow place value rules. As numbers get farther away from the decimal point to the left, the number gets larger. This large increase follows a specific pattern. Each place represents an additional value of 10. **Each time you move to the left, the number gets ten times bigger.**

7 = 7 ones
70 = 7 groups of 10 or 70 ones
700 = 7 groups of 100 or 70 groups of 10 or 700 ones

This pattern increases!

You can apply this idea when you multiple or divide by 10.

When you multiply by 10, you add an additional 10s place.

7 = 7 ones → **7 x 10 = 70** ones → Multiplying by 10 increases it to 7 groups of 10.
When you divide by 10, you remove an additional 10s place.

700 = 7 groups of **100** → **700 ÷ 10 = 70** → Dividing by 10 decreases it from 7 groups of 100 to 7 groups of 10.

1. 480 x 10 =

A. 48

B. 480

C. 4800

D. 48000

SHOW YOUR WORK

2. 20 ÷ 10 =

A. 2

B. 20

C. 200

D. 2000

SHOW YOUR WORK

3. 36 x 10 =

A. 3600

B. 360

C. 36

D. 36,000

SHOW YOUR WORK

4. How does 11,000 relate to 1, 100?

A. 1,100 is 10 times larger than 11,000.

B. 11,000 is 10 times smaller than 1,100.

C. 1,100 is 100 times smaller than 11,000.

D. 11,000 is 10 times larger than 1,100.

SHOW YOUR WORK

2.1.A | Generalize place value understanding for multi-digit whole numbers

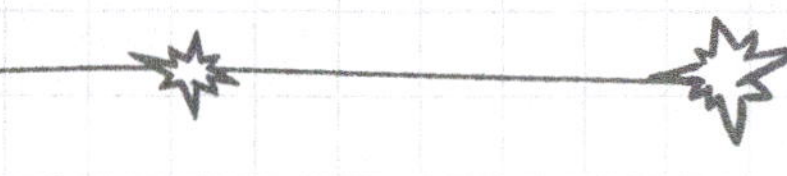

5. How would you increase the value of **600** to **6,000**? Explain your reasoning.

SHOW YOUR WORK

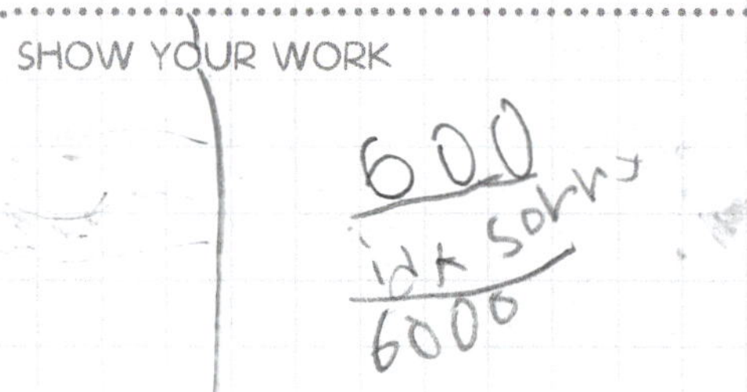

6. Which number is 10 times more than 380?

A. 38

B. 3810

C. 3800

D. 1038

SHOW YOUR WORK

7. Which number is 10 times less than 240?

A. 2400

B. 2410

C. 250

D. 24

SHOW YOUR WORK

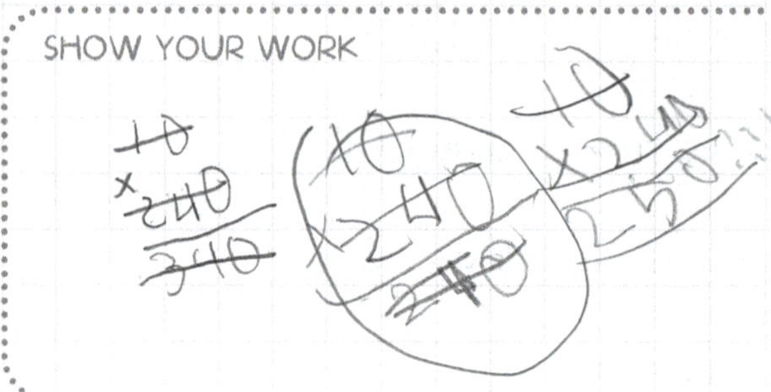

8. How does 720 compare to 72? Explain your reasoning.

SHOW YOUR WORK

9. How does 960 compare to 9,600?

A. It is ten times smaller.

B. It is ten times larger.

C. It is ten less.

D. It is ten more.

SHOW YOUR WORK

10. How would you decrease the value of 78,000 to 7,800? Explain your reasoning.

SHOW YOUR WORK

Now that we have reviewed the role of place value, let's examine the ways to read and write multi-digit numbers. Our number system is written in base ten form. When we write numbers such as 76 or 8,923, we are using base ten form.

Numbers can also be written as number names. You can think of this as using words to represent numbers. Using the example above, 76 becomes seventy-six and 8,923 becomes eight thousand, nine hundred, twenty-three.

Review the table below to remind yourself of how different values are represented by words.

Number	Words
1	Ones
10	Tens
100	Hundreds
1,000	Thousands
10,000	Ten-thousands
100,000	Hundred-thousands
1,000,000	Millions

When writing numbers with words, you should consider the place value as well as the numerical value of each place. Both need to be represented.

Another way to represent numbers is in expanded form. Expanded form breaks each number down into its separate place values. This can be helpful, especially when comparing numbers.

Look at our above examples, represented in expanded form:

76: 70 + 6
8,923: 8,000 + 900 + 20 + 3

Writing numbers in expanding form shows you exactly the value of each digit of each number.

This can be helpful when comparing numbers. When you compare numbers, you should start with the largest place value and compare those digits first. Then you work your way to the smallest place value.

Let's consider two numbers: 45 and 96.

Both numbers contain digits in the 10s and the 1s place. In the 10s place, you have a 4 and a 9. 9 is larger than 4, so 96 is larger than 45.

What if there is no value in a place?

For example, consider the numbers 1,025 and 1, 632.

Both numbers contain a 1 in the thousands place, so we move to the hundreds place. 6 is larger than 0, so 1,632 is larger than 1,025.

When you compare numbers, we use the following symbols:

< : Less than (means the number on the left is smaller than the number on the right)

> : Greater than (means the number on the left is larger than the number on the right)

≤ : Less than or equal to (means the number on the left is smaller than or equal to the number on the right)

≥ : Greater than or equal to (means the number on the left is larger than or equal to the number on the right)

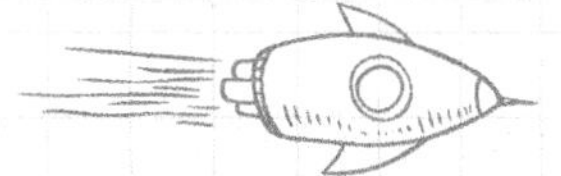

1. How would you represent three thousand, eight hundred, seven in base-ten form?

SHOW YOUR WORK

2. Which choice correctly represents 23, 094 as words?

A. Twenty-three thousand, ninety four

B. Twenty-three, ninety four

C. Twenty-three thousand, nine hundred four

D. Twenty-three, nine hundred four.

SHOW YOUR WORK

3. Which choice represents 729, 680 in proper expanded form?

A. 700,000 + 20,000 + 90,000+ 600 + 80

B. 700,000 + 29,000 + 600 + 8

C. 700,000 + 20,000 + 9,000 + 600 + 80

D. 729,000 + 680

SHOW YOUR WORK

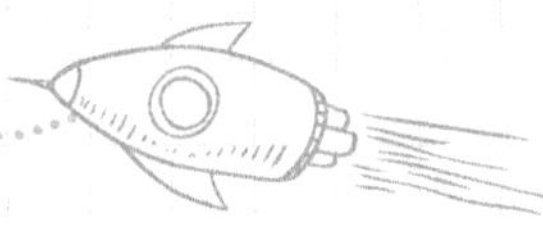

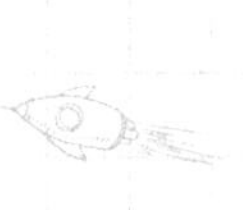

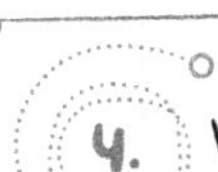

4. Which number is written in proper base-ten form?

A. 62, 000 + 368

B. 62,368

C. Sixty-two thousand, three hundred sixty-eight

D. 60,000 + 2,000 + 300 + 60 + 8

SHOW YOUR WORK

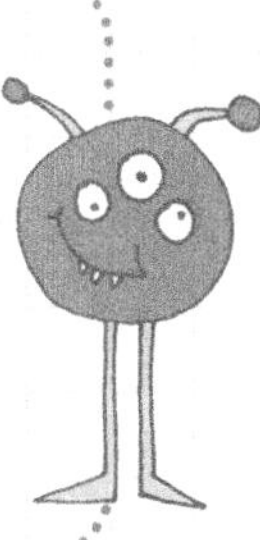

5. Which number equals eight hundred thousand, seven hundred twenty-nine?

A. 872, 900

B. 800, 709

C. 807, 029

D. 800, 729

SHOW YOUR WORK

6. Which number is larger: 729,326 or 792, 632? Explain your reasoning.

SHOW YOUR WORK

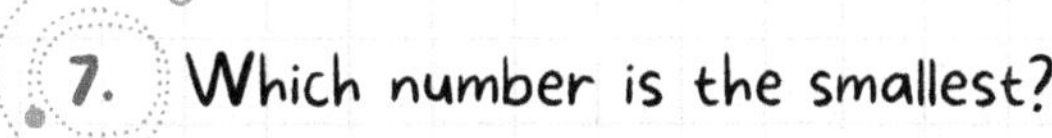

7. Which number is the smallest?

A. Three thousand, eight hundred fifty-two

B. Thirty thousand, five hundred eighty-two

C. Thirty-three thousand, eight hundred fifty-two

D. Three thousand, five hundred eighty-two

SHOW YOUR WORK

8. Which sign completes the comparison below?

70,000 + 8,000 + 900 + 60 + 8 ________ 80,000 + 7, 000 + 600 + 80 +9

A. >

B. <

C. ≥

D. ≤

SHOW YOUR WORK

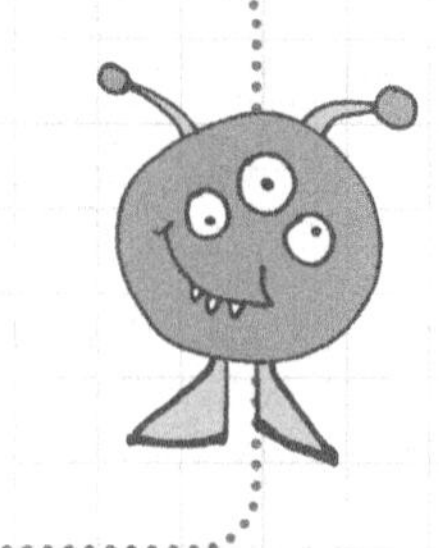

9. Which sign completes the comparison below?

850,625 ________ 805, 625

A. >

B. <

C. ≥

D. ≤

SHOW YOUR WORK

10. Which sign completes the comparison below?

Four hundred thousand, nine hundred eighty-two ____________ Forty thousand, nine hundred eight-two

A. $>$

B. $<$

C. $\geq$

D. $\leq$

SHOW YOUR WORK

NOTES

2.1.C | Generalize place value understanding for multi-digit whole numbers

Now that we have reviewed place value of numbers, let's talk about rounding. When you round a number, you are making the number simpler but keeping its value similar. Rounding can be helpful when estimating and solving problems and also when comparing large numbers. When you round, first you need to decide (or identify) which digit you are going to keep. You leave it the same if the next digit to the right is four or lower. If the next digit to the right is five or higher, you round the digit up.

Let's check out an example.

Round to the nearest thousands.

8, 256: (We're keeping the thousands place, so we review the hundreds place. **2** is lower than **4**, so thousands is going to stay the same. The number becomes **8,000**).

Round to the nearest thousands.

94, 873 (We're keeping the thousands place, so we review the hundreds place. **8** is larger than **5** so thousands is going to increase by **1**. The number becomes **95,000**).

1. Round to the nearest ten-thousand. 734,792. Explain your reasoning.

SHOW YOUR WORK

2. Round to the nearest hundred. 5, 932

A. 5,900

B. 6,000

C. 5, 800

D. 6,000

SHOW YOUR WORK

3. Round to the nearest thousand. 69, 872

A. 69,000

B. 70,000

C. 71,000

D. 9,000

SHOW YOUR WORK

4. Round to the nearest hundred-thousand. 312, 999

A. 312,000

B. 310,000

C. 400,000

D. 300,000

SHOW YOUR WORK

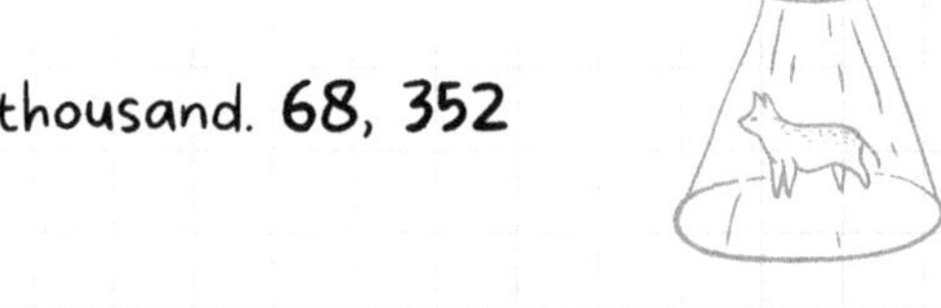

5. Round to the nearest ten-thousand. 68, 352

A. 60,000

B. 68,000

C. 70,000

D. 7,000

SHOW YOUR WORK

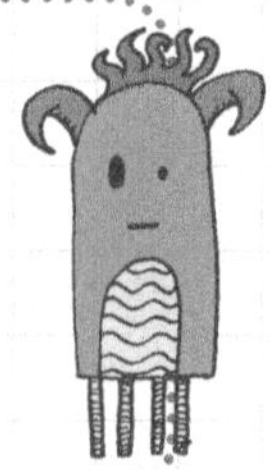

6. Last week, our library checked in 734 books. Round the number of books to the nearest hundred. Explain how you rounded 734 books to the nearest hundred.

SHOW YOUR WORK

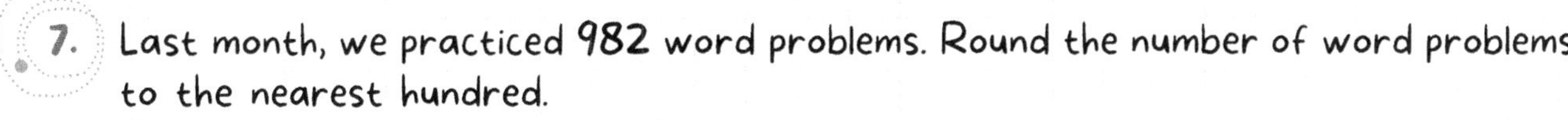

7. Last month, we practiced 982 word problems. Round the number of word problems to the nearest hundred.

A. 1,000
B. 900
C. 800
D. 980

SHOW YOUR WORK

8. We returned from a trip and told everyone we went around 1,100 miles. Which could be the exact length of our trip based on our rounded answer?

A. 1,193
B. 1,097
C. 1, 232
D. 986

SHOW YOUR WORK

9. Our school has around 1200 students.
What could be the exact number of students at our school?

A. 1298
B. 1254
C. 1183
D. 1134

SHOW YOUR WORK

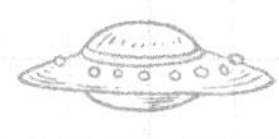

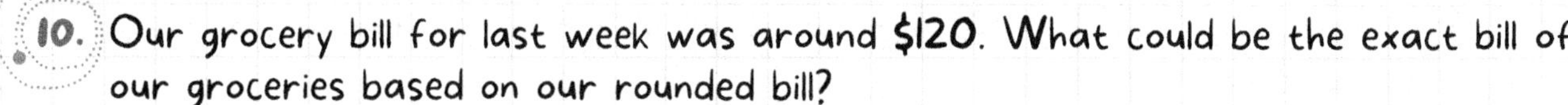

10. Our grocery bill for last week was around **$120**. What could be the exact bill of our groceries based on our rounded bill?

A. $ 125.73

B. $ 129.32

C. $ 114.45

D. $ 119.86

SHOW YOUR WORK

NOTES

2.2.A Use place value understanding and properties of operations to perform multi-digit arithmetic

Now that we have reviewed place value, we can apply those skills to add and subtract numbers. Remember, when we add numbers, we are combining amounts and when we are subtracting numbers, we are taking away from a total amount. When we add or subtract numbers, there are specific steps we follow. Let's review the steps using a set of numbers as example.

$$\begin{array}{r} 875 \\ +\ 729 \\ \hline \end{array}$$

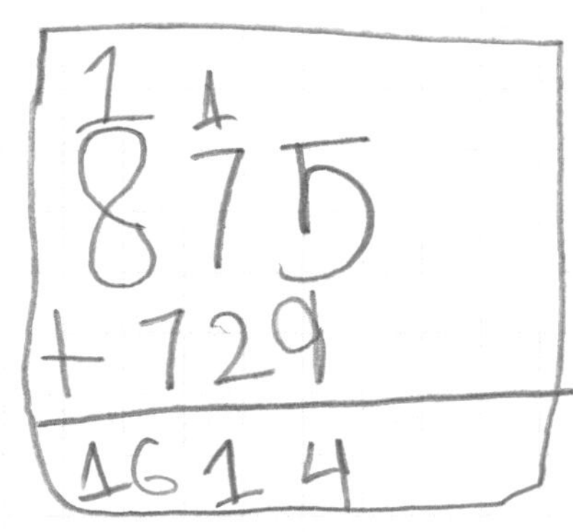

When we add, we start from the right and work left. We add each place value at a time and carry over any additional amounts to the next column.

Step 1: 5 + 9 = 13 We write down 3 under the 9 and then carry the 1 over to the tens column.

Step 2: 1 + 7 + 2 = 10 We write down 0 under the 2 and then carry the 1 over to the hundreds column.

Step 3: 1 + 8 + 7 = 16 We write down 16 as there are no more columns to add.

So, 875 + 729 = 1603

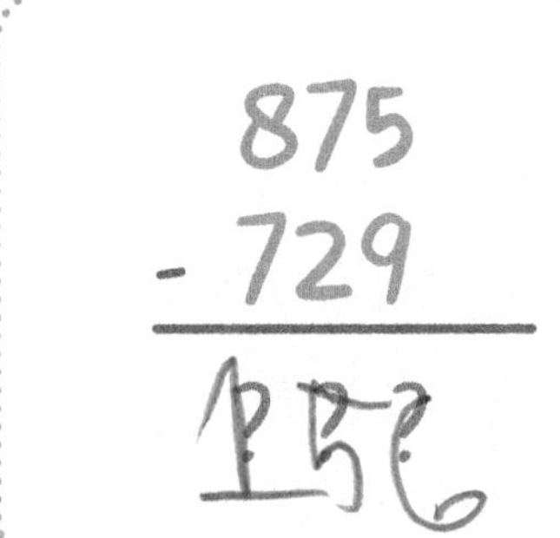

When we subtract, we also work from right to left. We subtract each place value one at a time and borrow from larger amounts as appropriate.

Step 1: 1 - 5 -9 cannot be done. We borrow a 10 from 7, so the 7 becomes 6 and the 5 becomes 15. 15 - 9 = 6

Step 1: 2 - 7 - 2 = 5

Step 1: 3 - 8 - 7 = 1

So, 875 - 729 = 156

2.2.A

Use place value understanding and properties of operations to perform multi-digit arithmetic

1. 682 + 2983 =

A. 5,663

B. 3,665

C. 3,556

D. 3,336

SHOW YOUR WORK

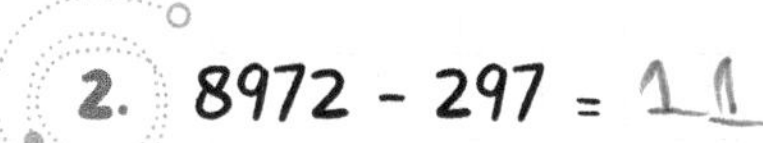

2. 8972 - 297 =

A. 7, 686

B. 7, 685

C. 8, 676

D. 8,675

SHOW YOUR WORK

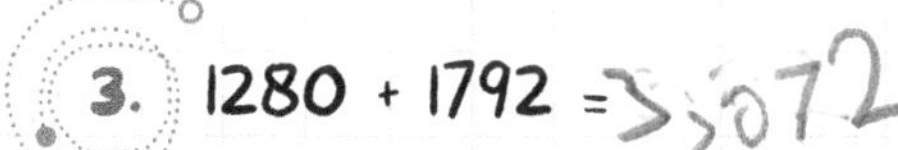

3. 1280 + 1792 =

A. 7, 032

B. 3,702

C. 3,072

D. 2, 372

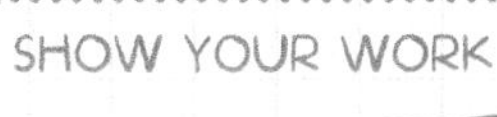

SHOW YOUR WORK

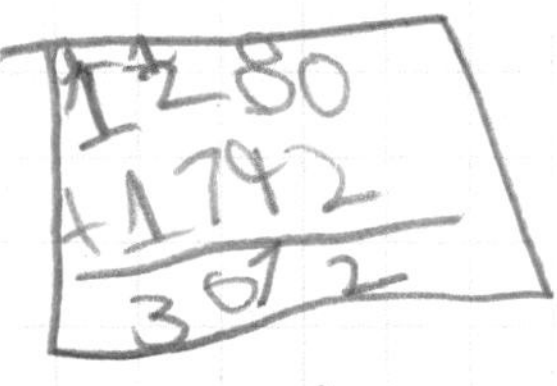

2.2.A

Use place value understanding and properties of operations to perform multi-digit arithmetic

4. 920 - 293 = 721

A. 627

B. 726

C. 727

D. 626

SHOW YOUR WORK

5. 23082 + 3237 =

A. 19,319

B. 27,319

C. 19,845

D. 26,319

SHOW YOUR WORK

6. 8129 - 201 = 7928

A. 8,927

B. 8,278

C. 7,928

D. 7,298

SHOW YOUR WORK

ARGOPREP

2.2.A

Use place value understanding and properties of operations to perform multi-digit arithmetic

7. 3972 + 3789 =

A. 9,872

B. 7,761

C. 7,828

D. 7,982

SHOW YOUR WORK

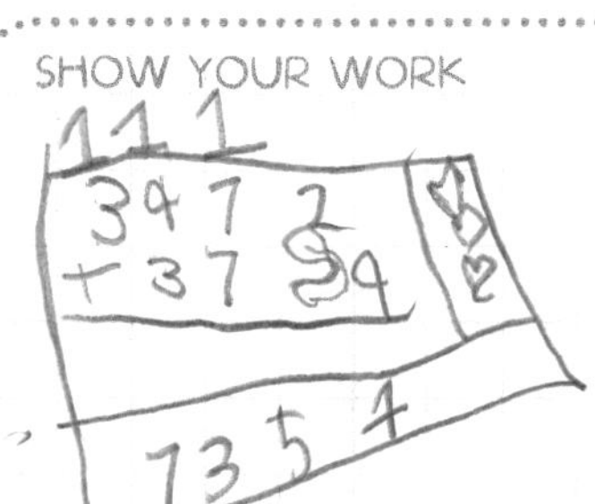

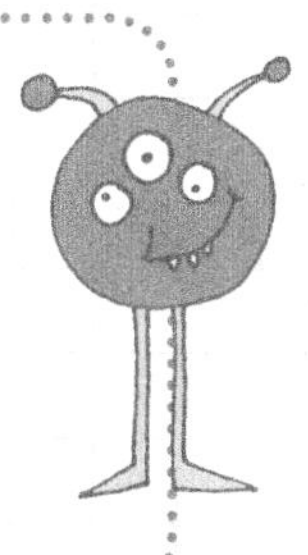

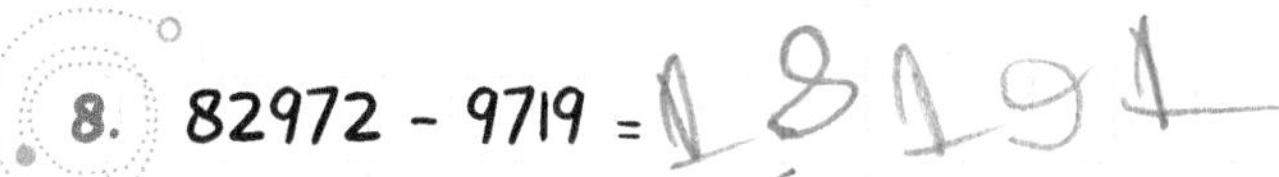

8. 82972 - 9719 =

A. 72,533

B. 73,253

C. 72,523

D. 73,552

SHOW YOUR WORK

9. 982 + 334 =

A. 1,316

B. 1,136

C. 1,313

D. 632

SHOW YOUR WORK

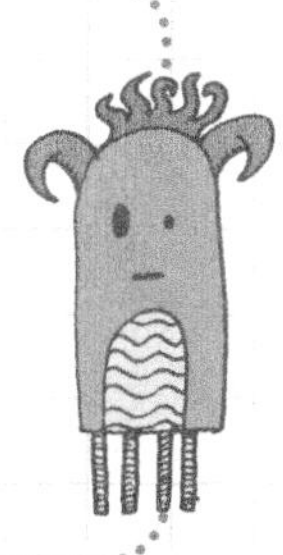

2.2.A

Use place value understanding and properties of operations to perform multi-digit arithmetic

10. 678 - 234 =

A. 688

B. 422

C. 33

D. 444

SHOW YOUR WORK

NOTES

2.2.B Use place value understanding and properties of operations to perform multi-digit arithmetic

Now that we have mastered addition and subtraction, let's revisit multiplication. We have already reviewed multiplication of single digit numbers, we can apply those skills as well as our skills of addition and subtraction to multiply numbers with more than one digit. Let's look closer at how to do this!

We can break multiplication problems into smaller problems to solve them. Let's review how to do this with the problem 37 x 8.

We learned how 37 can be represented as 30 + 7. We can use that information to make 37 x 8 a more manageable problem.

First, break 37 into its expanded form. Then, multiply each part of its expanded form by 8. Finally add each of the results together to get your final result.

37 x 8 = 30 x 8 = 240
+ 7 x 8 = 56

37 x 8 = 240 + 56 = 296

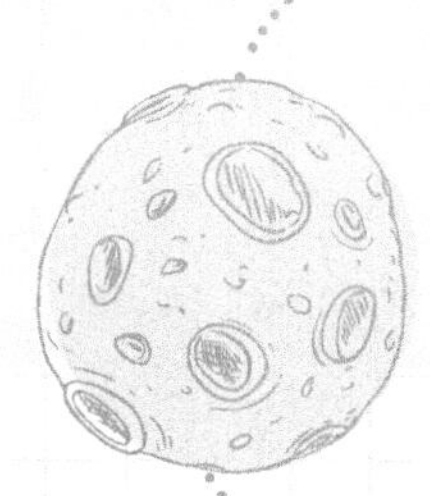

It works with 2 digit multiplication too!

82 x 15 = 80 + 2 x 10 + 5

We multiply 80 by 10 and by 5 to get 800 and 400.
Then we multiply 2 by 10 and by 5 to get 20 and 10.
We add all those products together (800 + 400 + 20 + 10) for our answer of 1230.

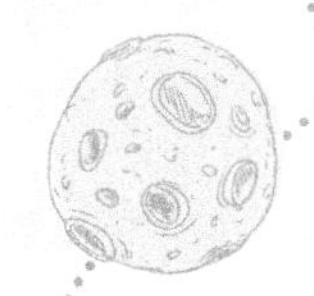

By breaking it down this way, we are able to solve the problem quickly.

2.2.B

Use place value understanding and properties of operations to perform multi-digit arithmetic

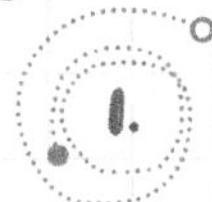

1. 739 x 2 =

A. 1,748

B. 1,487

C. 1,478

D. 1,847

SHOW YOUR WORK

2. 92 x 31 =

A. 2,852

B. 2,582

C. 2,722

D. 2,654

SHOW YOUR WORK

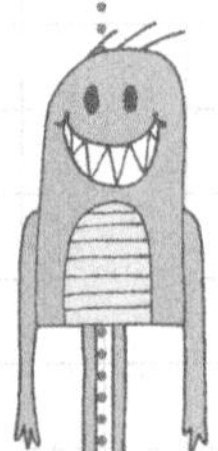

3. 678 x 4 =

A. 2,172

B. 2,712

C. 2,721

D. 2,271

SHOW YOUR WORK

4. 75 x 82 =

A. 5,601

B. 5,610

C. 6,510

D. 6,150

SHOW YOUR WORK

5. 1458 x 8 =

A. 11,614

B. 11,664

C. 11,416

D. 11,466

SHOW YOUR WORK

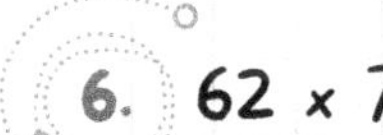

6. 62 x 74 =

A. 5,884

B. 5,488

C. 4,588

D. 4,858

SHOW YOUR WORK

2.2.B

Use place value understanding and properties of operations to perform multi-digit arithmetic

7. 72 x 32 =

A. 2,304

B. 2,403

C. 2,034

D. 2,043

SHOW YOUR WORK

8. 54 x 62 =

A. 8,433

B. 7,328

C. 2,438

D. 3,348

SHOW YOUR WORK

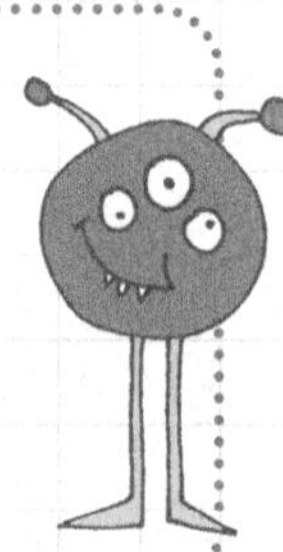

9. 7125 x 6 =

A. 42,750

B. 47,250

C. 45,270

D. 42,705

SHOW YOUR WORK

2.2.B

Use place value understanding and properties of operations to perform multi-digit arithmetic

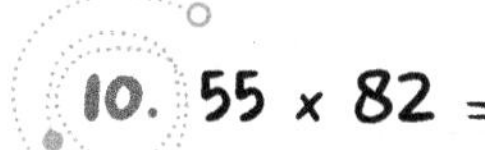

10. 55 x 82 =

A. 5,410

B. 4,510

C. 1,045

D. 1,540

SHOW YOUR WORK

NOTES

2.2.C

Use place value understanding and properties of operations to perform multi-digit arithmetic

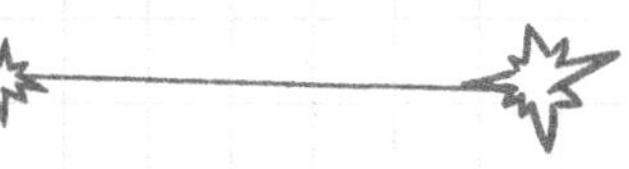

Now that we have mastered multiplication, let's review division! In division, we need to consider each digit carefully as our problem.

Let's look at the problem 272 ÷ 4.

First, we need to set the problem up so we can divide it properly.

$$4 \overline{)\,272}$$

We start with the largest place value and work our way to the smallest.

First we ask ourselves, does 4 go into 2 evenly? The answer is no, so we consider 4 compared to 27. 4 goes into 27, 6 times. We then take 6 x 4, which is 24- and write it under 27.

```
      6
   ______
4 |  272
   - 24
      3
```

27 - 24 = 3 we need to bring down the 2, so the number becomes 32

```
      6
   ______
4 |  272
   - 24
      32
```

How many times does 4 go into 32? 8 x 4 is 32, so that works perfectly!

```
     68
   ┌────
 4 │ 272
   - 24
      32
    - 32
       0
```

There is no remainder for this problem.

Let's look at another:

```
   ┌────
 6 │ 423
```

How many times does 6 go into 4? None, so we need to consider 6 into 42.

```
     7
   ┌────
 6 │ 423
   - 42
      0
```

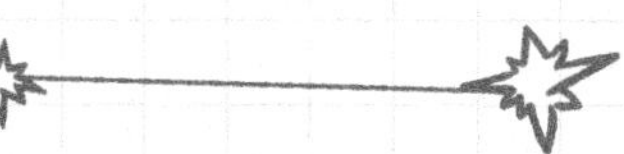

6 x 7 is 42, so that goes under the 42 and we subtract. We end up with 0 and we bring down the 3.

```
     7
6 | 423
   -42
    03
```

How many times does 6 go into 3? 0! 0x 6 = 0 so we end up with a remainder of 3 in this problem, which we express as R3.

```
     70  R3
6 | 423
   -42
    03
   - 0
     3
```

2.2.C

Use place value understanding and properties of operations to perform multi-digit arithmetic

1. 225 ÷ 3 =

A. 65

B. 55

C. 45

D. 75

SHOW YOUR WORK

2. 147 ÷ 7 =

A. 21

B. 20

C. 22

D. 23

SHOW YOUR WORK

3. 865 ÷ 5 =

A. 172

B. 173

C. 174

D. 170

SHOW YOUR WORK

4. 312 ÷ 6 =

A. 56

B. 50

C. 52

D. 54

SHOW YOUR WORK

5. 3927 ÷ 7 =

A. 523

B. 561

C. 563

D. 536

SHOW YOUR WORK

6. 146 ÷ 3 =

A. 48 R2

B. 48

C. 48 R1

D. 42

SHOW YOUR WORK

2.2.C

Use place value understanding and properties of operations to perform multi-digit arithmetic

7. 853 ÷ 4 =

A. 213

B. 213 R1

C. 214 R1

D. 213 R2

SHOW YOUR WORK

8. 8352 ÷ 6 =

A. 1923

B. 1239

C. 1392

D. 1293

SHOW YOUR WORK

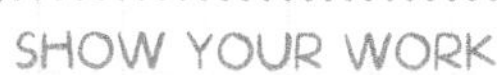

9. 528 ÷ 8 =

A. 63 R3

B. 64

C. 65

D. 66

SHOW YOUR WORK

10. 786 ÷ 5 =

A. 157 R1

B. 157

C. 158 R2

D. 158

SHOW YOUR WORK

NOTES

2.3. Chapter Test

1. How does 34,000 compare to 3,400?

A. Ten times larger

B. Ten times smaller

C. Ten more

D. Ten less

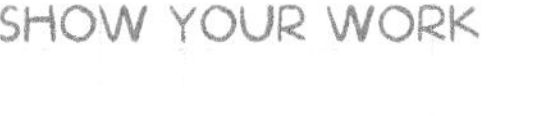

SHOW YOUR WORK

2. Which number is ten times larger than 7,800?

A. 7,810

B. 78,000

C. 780

D. 7,900

SHOW YOUR WORK

3. Which number is largest?

A. 962, 328

B. 992, 382

C. 992, 385

D. 962, 832

SHOW YOUR WORK

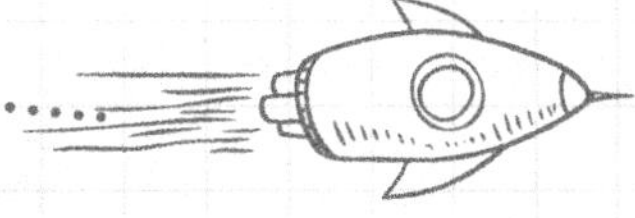

2.3. Chapter Test

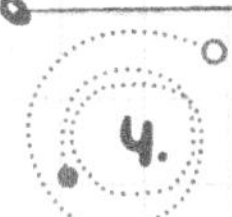

4. Which number is smallest?

A. Thirty-thousand, seven hundred ninety-five

B. Thirty-thousand, seventy-nine

C. Thirty thousand, seven hundred fifty-nine

D. Thirty thousand, ninety-five

SHOW YOUR WORK

5. Which sign completes the comparison below?

800,000 + 9,000 + 90 +8 ____________ 800,000 + 90,000 + 800 +9

A. >

B. <

C. ≥

D. ≤

SHOW YOUR WORK

6. Round to the nearest thousand. 81, 632

A. 82,000

B. 81,000

C. 80,000

D. 81, 600

SHOW YOUR WORK

2.3. Chapter Test

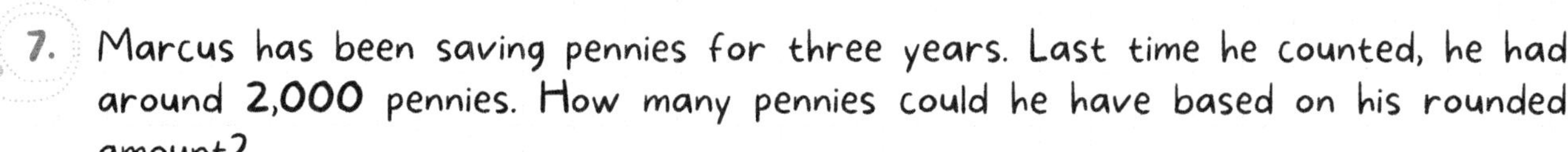

7. Marcus has been saving pennies for three years. Last time he counted, he had around **2,000** pennies. How many pennies could he have based on his rounded amount?

A. 2, 562

B. 2,486

C. 2, 893

D. 1,422

8. Our family car traveled around **200,000** miles before we got a new one. Which choice could be the possible mileage based on our rounded amount?

A. 143, 879

B. 254, 681

C. 212,868

D. 145,689

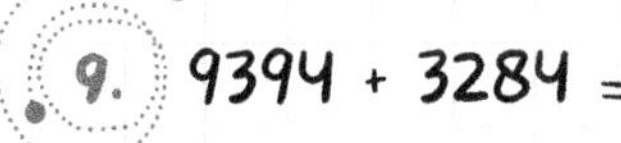

9. 9394 + 3284 =

A. 11, 678

B. 12,876

C. 21, 768

D. 12,678

10. 7292 - 208 =

A. 7, 480

B. 8,740

C. 7,804

D. 7,084

SHOW YOUR WORK

11. 643 x 6 =

A. 3,858

B. 3,588

C. 5,388

D. 3,885

12. 29 x 31 =

A. 601

B. 899

C. 989

D. 901

2.3. Chapter Test

13. 1427 × 8 =

A. 11,614

B. 16,141

C. 11,416

D. 11,146

14. 2144 ÷ 4 =

A. 653

B. 635

C. 563

D. 536

15. 523 ÷ 2 =

A. 261 R1

B. 261

C. 262 R2

D. 621

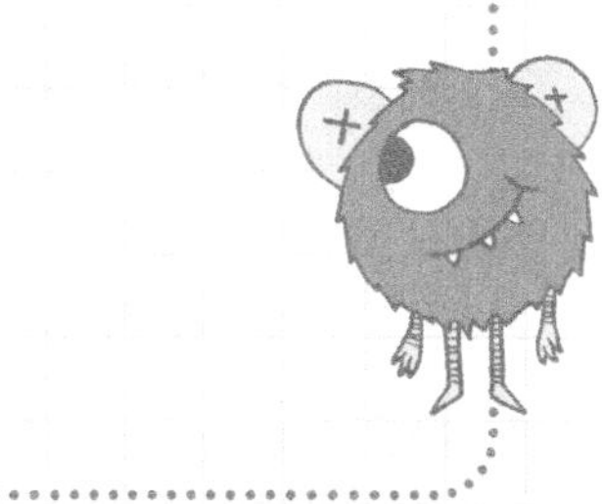

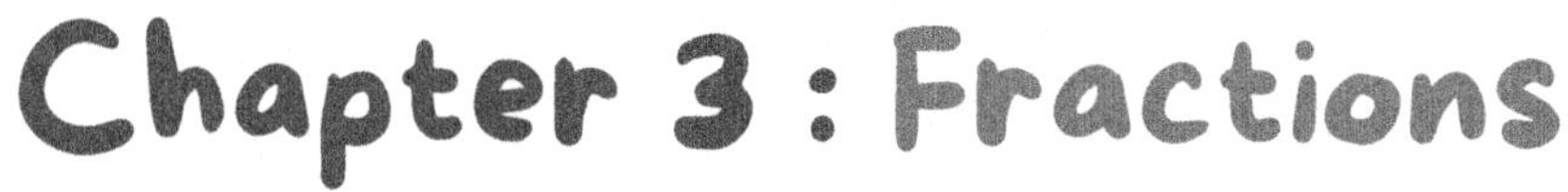
Chapter 3 : Fractions

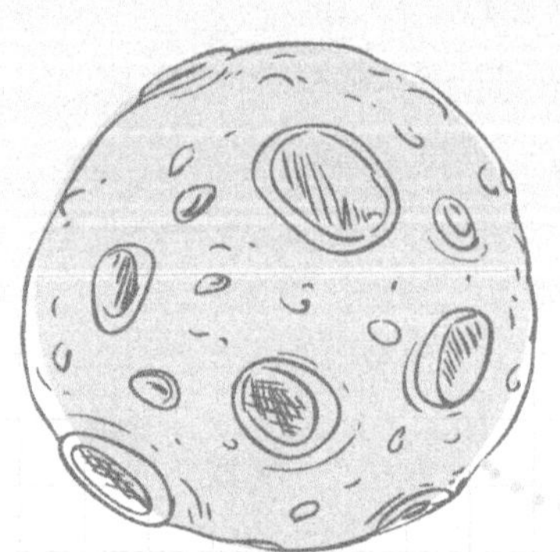

ARGOPREP

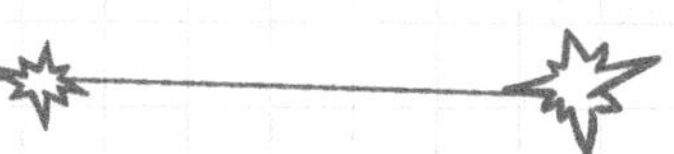

Just as we can express equivalent numbers using powers, we can express equivalent fractions. **Remember, a fraction is composed of two parts: the numerator and the denominator. Fractions are expressed as a part of the whole, so the numerator on top is the part, and the denominator on the bottom is the whole.**

For example in the fraction $\frac{1}{2}$, the numerator is 1 and the denominator is 2. The fraction $\frac{1}{2}$ expresses an amount that is comprised as 1 part out of two parts.

Fractions can be illustrated as shapes broken into smaller parts. It is always possible to break a shape into smaller parts. Consider the shape below.

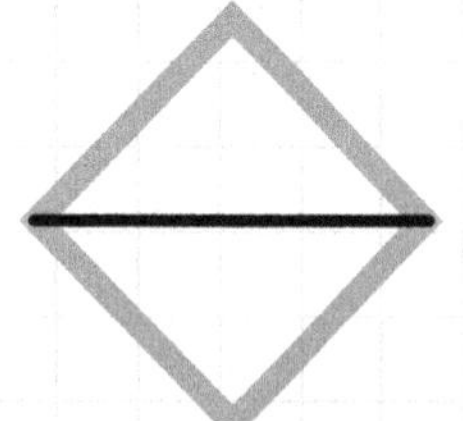

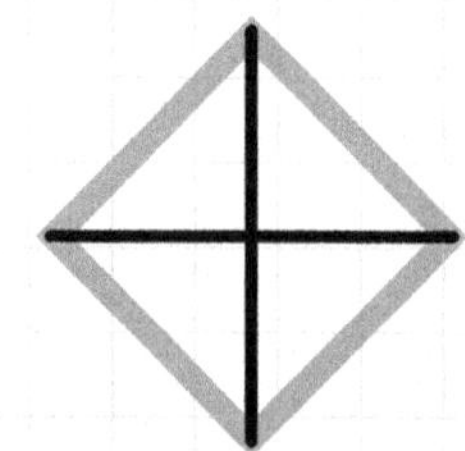

We can color the same amount on each diamond. If we color one part of the first diamond, it is equivalent to two parts on the second diamond and also 4 parts on the third diamond. Those would all be different fractions ($\frac{1}{2}$, $\frac{2}{4}$ and $\frac{4}{8}$) but they all take up the same amount of the diamond.

To get from one equivalent fraction to another, you just need to do the same thing to both parts of the fraction. Using the example above, to go from $\frac{1}{2}$ to $\frac{2}{4}$, you need to multiply both the numerator and denominator by 2. You can do it again to go from $\frac{2}{4}$ to $\frac{4}{8}$. You can also go from $\frac{1}{2}$ to $\frac{4}{8}$ by multiplying both the numerator and denominator by 4. **Any number works as long as you can do it to both the numerator and denominator.**

3.1.A | Extend understanding of fraction equivalence and ordering.

1. Which fraction is equivalent to $\frac{4}{7}$?

A. $\frac{12}{21}$

B. $\frac{12}{7}$

C. $\frac{4}{21}$

D. $\frac{12}{28}$

SHOW YOUR WORK

2. Which fraction is equivalent to $\frac{3}{9}$?

A. $\frac{1}{2}$

B. $\frac{1}{3}$

C. $\frac{1}{4}$

D. $\frac{1}{9}$

SHOW YOUR WORK

3. Which fraction is equivalent to $\frac{1}{6}$?

A. $\frac{4}{6}$

B. $\frac{2}{6}$

C. $\frac{3}{12}$

D. $\frac{4}{24}$

SHOW YOUR WORK

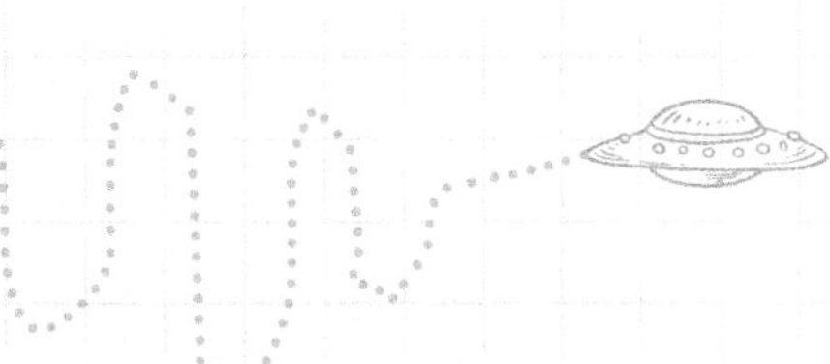

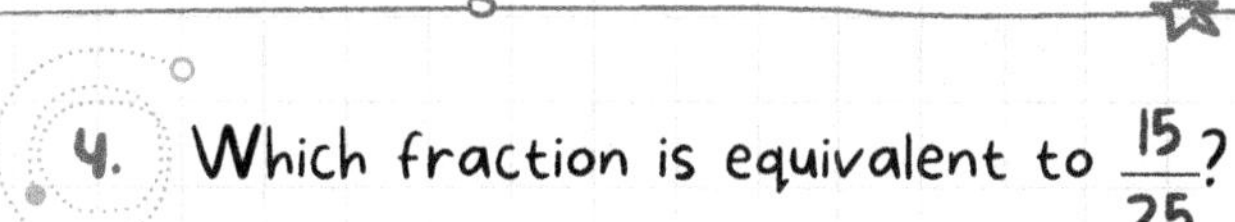

4. Which fraction is equivalent to $\frac{15}{25}$?

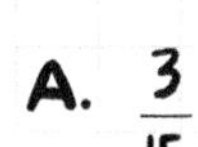

A. $\frac{3}{15}$

B. $\frac{1}{2}$

C. $\frac{3}{5}$

D. $\frac{5}{3}$

SHOW YOUR WORK

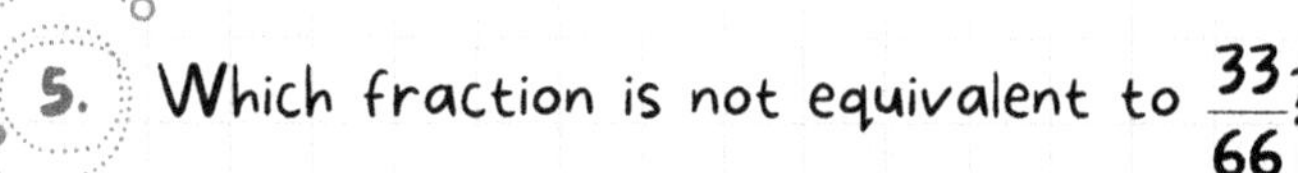

5. Which fraction is not equivalent to $\frac{33}{66}$?

A. $\frac{2}{3}$

B. $\frac{3}{6}$

C. $\frac{1}{2}$

D. $\frac{11}{22}$

SHOW YOUR WORK

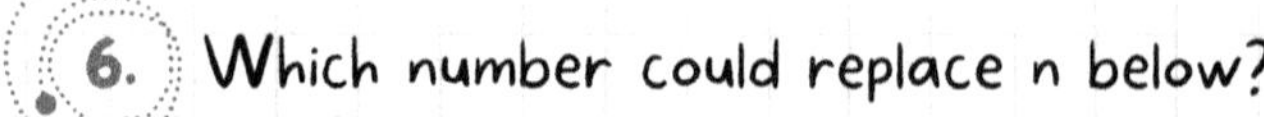

6. Which number could replace n below?

$$\frac{n}{7} = \frac{16}{28}$$

A. 4

B. 5

C. 6

D. 7

SHOW YOUR WORK

7. Which number could replace d below?

$$\frac{6}{d} = \frac{2}{12}$$

A. 12

B. 3

C. 18

D. 36

SHOW YOUR WORK

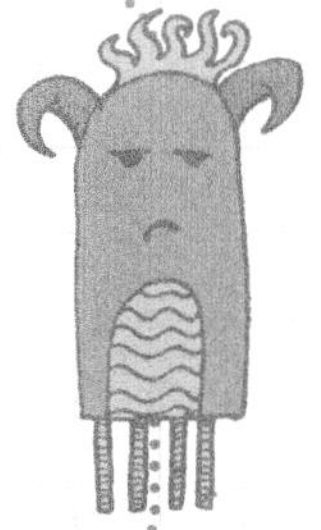

8. Which number could replace n below?

$$\frac{12}{18} = \frac{n}{6}$$

A. 2

B. 3

C. 4

D. 5

SHOW YOUR WORK

9. Which number could replace d below?

$$\frac{2}{d} = \frac{18}{81}$$

A. 3

B. 9

C. 12

D. 15

SHOW YOUR WORK

10. Divide the rectangle into another set of shapes equivalent to the one below:

SHOW YOUR WORK

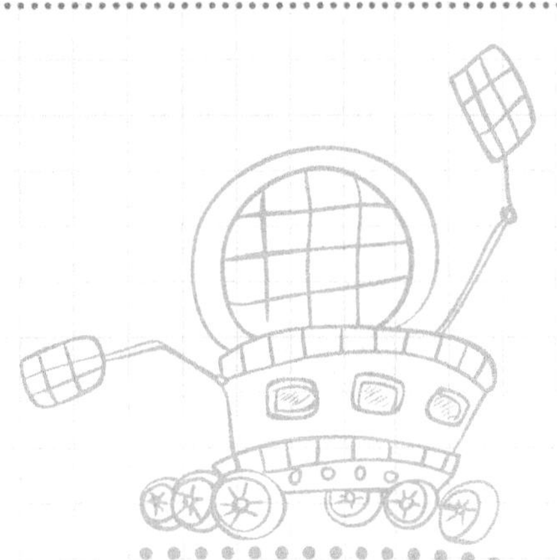

3.1.B | Extend understanding of fraction equivalence and ordering.

We reviewed how to compare large numbers. Now let's start to examine how to compare fractions. Comparing fractions uses the same symbols as comparing large numbers:

$<$: Less than (means the number on the left is smaller than the number on the right)

$>$: Greater than (means the number on the left is larger than the number on the right)

$\leq$: Less than or equal to (means the number on the left is smaller than or equal to the number on the right)

$\geq$: Greater than or equal to (means the number on the left is larger than or equal to the number on the right)

The best way to compare fractions is to use equivalent denominators. Once you have fractions with like denominators, it is easy to see how the two fractions compare.

Let's look at an example.

$\frac{1}{4}$ ------ $\frac{2}{6}$

First, what number do 4 and 6 have in common? They both are factors of 12! So we rewrite the fractions as

$\frac{1}{4} = \frac{3}{12}$ and $\frac{2}{6} = \frac{4}{12}$

Then, rewrite the comparison as $\frac{3}{12}$ ------ $\frac{4}{12}$.

The correct sign for comparison is $<$, because 3 is smaller than 4.

3.1.B | Extend understanding of fraction equivalence and ordering.

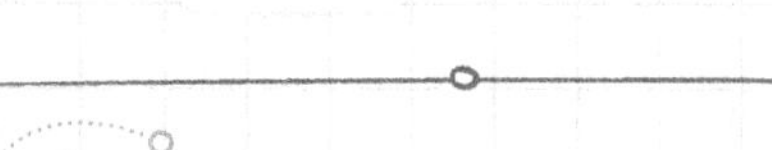

1. When comparing $\frac{1}{8}$ and $\frac{3}{12}$, which common denominator should you use?

A. 8

C. 12

B. 24

D. 4

How do you know?

SHOW YOUR WORK

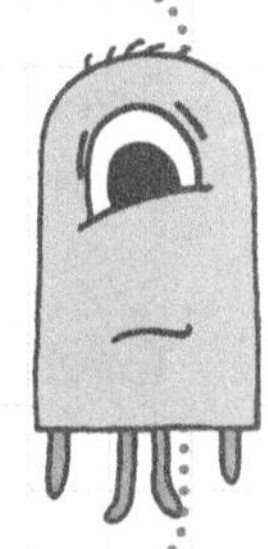

2. When comparing $\frac{3}{7}$ and $\frac{2}{3}$, which common denominator should you use?

A. 7

C. 21

B. 14

D. 28

How do you know?

SHOW YOUR WORK

3. When comparing $\frac{2}{6}$ and $\frac{1}{3}$, which common denominator should you use?

A. 6

B. 12

C. 15

D. 18

SHOW YOUR WORK

4. When comparing $\frac{3}{5}$ and $\frac{4}{6}$, which common denominator should you use?

A. 12

B. 10

C. 15

D. 30

SHOW YOUR WORK

5. When comparing $\frac{5}{12}$ and $\frac{4}{8}$, which common denominator should you use?

A. 4

B. 12

C. 24

D. 16

SHOW YOUR WORK

6. $\frac{2}{3}$ ------ $\frac{6}{12}$

A. $>$

B. $<$

C. $=$

D. $\geq$

SHOW YOUR WORK

7. $\frac{4}{7}$ ------ $\frac{12}{18}$

A. $>$

B. $<$

C. $=$

D. $\geq$

SHOW YOUR WORK

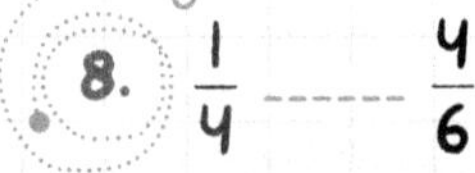

8. $\frac{1}{4}$ ------ $\frac{4}{6}$

A. $>$

B. $<$

C. $=$

D. $\geq$

SHOW YOUR WORK

9. $\frac{7}{8}$ ------ $\frac{8}{9}$

A. $>$

B. $<$

C. $=$

D. $\geq$

SHOW YOUR WORK

10. $\frac{3}{5}$ ------ $\frac{2}{7}$

A. $>$

B. $<$

C. $=$

D. $\geq$

SHOW YOUR WORK

NOTES

3.2.A

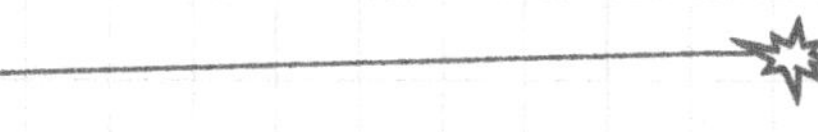

Build fractions from unit fractions

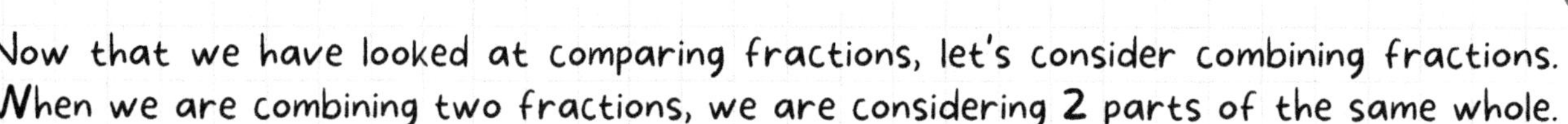

Now that we have looked at comparing fractions, let's consider combining fractions. When we are combining two fractions, we are considering 2 parts of the same whole.

Let's think about it real world terms. If you eat $\frac{2}{8}$ slices of a pizza and your brother eats $\frac{3}{8}$ slices of pizza, you only have $\frac{3}{8}$ of a pizza left. You are not adding more slices to the pizza when you add fractions but you are combining more parts of the whole.

Let's look at another example:

I have an apple and I divide it into 8 equal parts. Each part of the apple is $\frac{1}{8}$. I eat three of the parts or $\frac{3}{8}$. My sister eats two of the parts or $\frac{2}{8}$. How much of the apple did we eat?

2 parts+ 3 parts = 5 parts so we ate $\frac{5}{8}$ of the apple.

It works the same with subtraction too!

3.2.A | Build fractions from unit fractions

1. $\frac{2}{8} + \frac{3}{8} =$

A. $\frac{6}{24}$

B. $\frac{6}{64}$

C. $\frac{5}{16}$

D. $\frac{5}{8}$

SHOW YOUR WORK

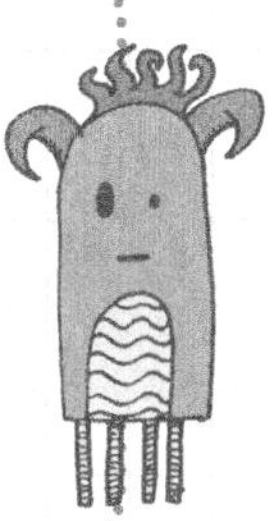

2. $\frac{4}{7} + \frac{1}{7} =$

A. $\frac{4}{49}$

B. $\frac{4}{14}$

C. $\frac{5}{7}$

D. $\frac{5}{14}$

SHOW YOUR WORK

3. $\frac{6}{16} - \frac{2}{16} =$

A. $\frac{4}{0}$

B. $\frac{4}{16}$

C. $\frac{4}{8}$

D. $\frac{2}{4}$

SHOW YOUR WORK

4. There are 9 buttons on my shirt. Four buttons fell off. How many buttons are left?

A. 2

B. 3

C. 4

D. 5

SHOW YOUR WORK

5. Our family had pie for dessert. Each member of our family had 1 piece. If there are 6 members of our family and the pie was divided into 8 pieces, write a problem that represents how many pieces were eaten.

SHOW YOUR WORK

NOTES

3.2.B | Build fractions from unit fractions

Something else we should consider about fractions is their smaller components. **Let's review our pizza example again.** If we break a pizza into 8 slices, each slice is $\frac{1}{8}$ of the pizza. If we take 3 slices, we have taken $\frac{1}{8} + \frac{1}{8} + \frac{1}{8}$ or $\frac{3}{8}$ slices. We can also do this with mixed numbers. If we have 3 pizzas of 8 slices each, we have 3 pizzas with $\frac{8}{8}$ slices. If we start to eat pizza and take 4 slices from one pizza, we still have $\frac{8}{8} + \frac{8}{8} + \frac{5}{8}$ to feed our friends. This is the same as representing it as 2 whole pizzas and $\frac{5}{8}$ of the other pizza or $2\frac{5}{8}$.

1. Which fraction problem is equivalent to $\frac{7}{8}$?

A. $\frac{3}{8} + \frac{4}{8}$

B. $\frac{1}{8} + \frac{7}{8}$

C. $\frac{1}{7} + \frac{1}{8}$

D. $\frac{7}{1} + \frac{8}{1}$

SHOW YOUR WORK

2. Which fraction problem is equivalent to $\frac{9}{12}$?

A. $\frac{3}{12} + \frac{3}{12}$

B. $\frac{6}{12} + \frac{1}{12}$

C. $\frac{3}{12} + \frac{3}{12} + \frac{3}{12}$

D. $\frac{9}{12} + \frac{1}{12}$

SHOW YOUR WORK

3. Which fraction problem is equivalent to $2\frac{3}{4}$?

A. $\frac{4}{4} + \frac{3}{4}$

B. $\frac{4}{4} + \frac{4}{4} + \frac{3}{4}$

C. $\frac{5}{4} + \frac{4}{4}$

D. $\frac{1}{4} + \frac{3}{4}$

SHOW YOUR WORK

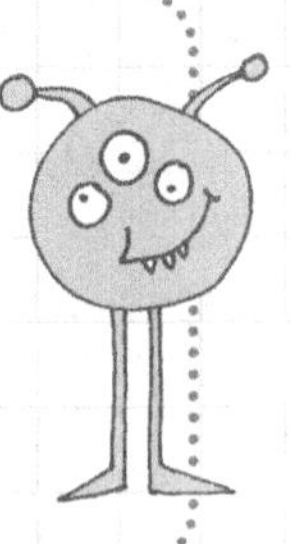

4. Which fraction problem is not equivalent to $\frac{5}{6}$?

A. $\frac{2}{6} + \frac{3}{6}$

B. $\frac{2}{6} + \frac{2}{6} + \frac{1}{6}$

C. $\frac{1}{6} + \frac{1}{6} + \frac{1}{6} + \frac{2}{6}$

D. $\frac{5}{6} + \frac{1}{6}$

SHOW YOUR WORK

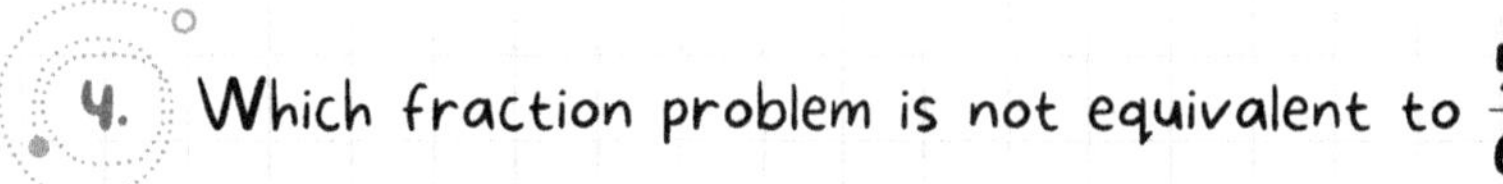

5. Which fraction problem is not equivalent to $3\frac{3}{4}$?

A. $\frac{4}{3} + \frac{3}{3}$

B. $\frac{4}{4} + \frac{4}{4} + \frac{4}{4} + \frac{3}{4}$

C. $\frac{8}{4} + \frac{7}{4}$

D. $\frac{1}{4} + \frac{1}{4} + \frac{1}{4} + \frac{8}{4} + \frac{4}{4}$

SHOW YOUR WORK

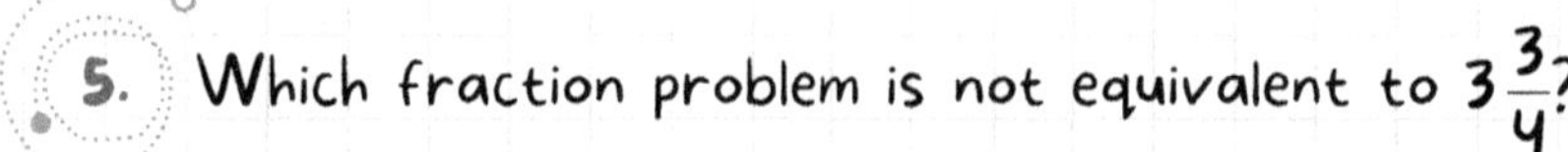

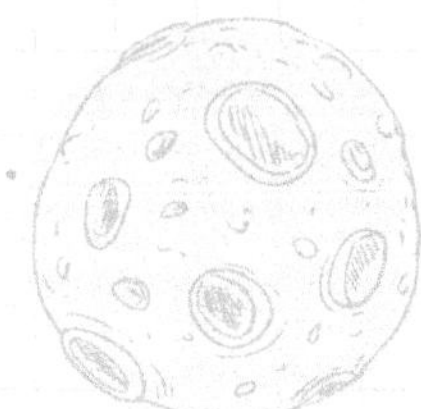

3.2.C | Build fractions from unit fractions

When we combine mixed numbers with addition and subtraction, we need to follow the same rules as when we combine whole numbers and fractions with addition and subtraction. One way to solve a mixed number problem is to rewrite the problem as two smaller problems, the mixed number operation and the fraction operation.

For example $4\frac{1}{4} + 5\frac{2}{4} = 4 + 5$ AND $\frac{1}{4} + \frac{2}{4}$

First, we can add $4 + 5 = 9$

Then, we can add $\frac{1}{4} + \frac{2}{4} = \frac{3}{4}$

You combine the answers for the final answer: $9\frac{3}{4}$

Let's look at subtraction: $5\frac{1}{3} - 2\frac{2}{3}$

First, rewrite the problem as $5 - 2$ AND $\frac{1}{3} - \frac{2}{3}$.

Now, we can't take $\frac{1}{3} - \frac{2}{3}$ so we rewrite it again as $4 - 2$ and $\frac{4}{3} - \frac{2}{3}$.

Then we complete the operations.

First $4 - 2 = 2$

Then, $\frac{4}{3} - \frac{2}{3} = \frac{2}{3}$

You combine the answers for the final answer $2\frac{2}{3}$.

1. $1\frac{3}{7} + 3\frac{4}{7} =$

A. $4\frac{4}{7}$

B. 5

C. $3\frac{7}{7}$

D. $\frac{7}{7}$

SHOW YOUR WORK

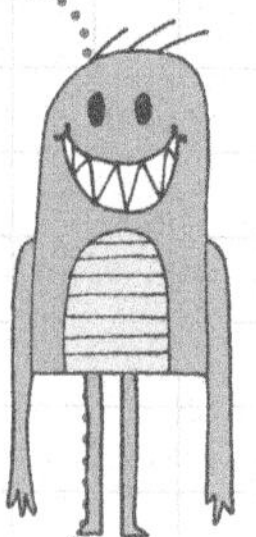

2. $2\frac{6}{9} - 1\frac{2}{9} =$

A. $1\frac{3}{9}$

B. $3\frac{8}{9}$

C. $1\frac{4}{9}$

D. $\frac{4}{9}$

SHOW YOUR WORK

3. $3\frac{1}{8} + 4\frac{5}{8} =$

A. $7\frac{6}{8}$

B. $8\frac{6}{8}$

C. $8\frac{1}{8}$

D. $7\frac{5}{8}$

SHOW YOUR WORK

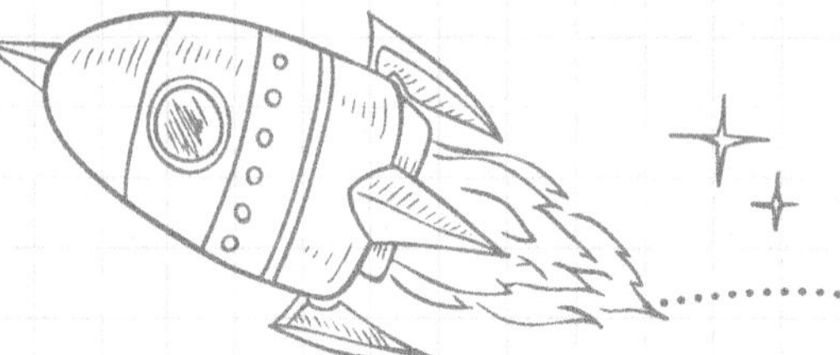

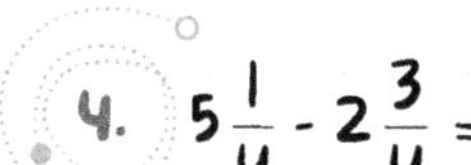

4. $5\frac{1}{4} - 2\frac{3}{4} =$

A. $3\frac{1}{4}$

B. $2\frac{3}{4}$

C. $3\frac{3}{4}$

D. $2\frac{2}{4}$

SHOW YOUR WORK

5. $2\frac{3}{4} + 5\frac{1}{4} =$

A. 8

B. $7\frac{3}{4}$

C. $\frac{8}{4}$

D. $8\frac{1}{4}$

SHOW YOUR WORK

NOTES

3.2.D | Build fractions from unit fractions

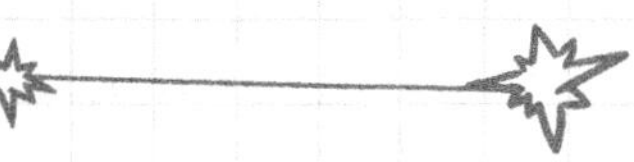

Do you know how to apply these skills in word problems? Remember the following steps when you are trying to solve word problems:

1 → Read the word problem.

2 → Reread the problem and circle what you want to know.

3 → Reread the problem and underline what you know.

4 → Write a problem based on what you know.

5 → Apply the skills you have learned so far to solve the problem and check your answer.

Remember to phrase your answer in terms of the real world.

1. Sydney practices piano every day for $\frac{3}{4}$ of an hour. After three days, how long has she practiced?

A. $\frac{3}{4}$

B. $\frac{9}{4}$

C. $\frac{6}{4}$

D. $\frac{9}{12}$

SHOW YOUR WORK

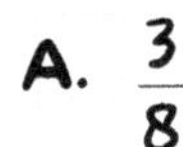

2. Isaac had to read a book in a week. The book has 8 chapters. He read 2 chapters today and 3 chapters yesterday. How much of the book has he read total?

A. $\frac{3}{8}$

B. $\frac{2}{8}$

C. $\frac{5}{8}$

D. $\frac{8}{5}$

3. Using the answer to the problem above, how much of the book does he have left to read?

A. $\frac{3}{8}$

B. $\frac{5}{8}$

C. $\frac{8}{8}$

D. $\frac{1}{8}$

SHOW YOUR WORK

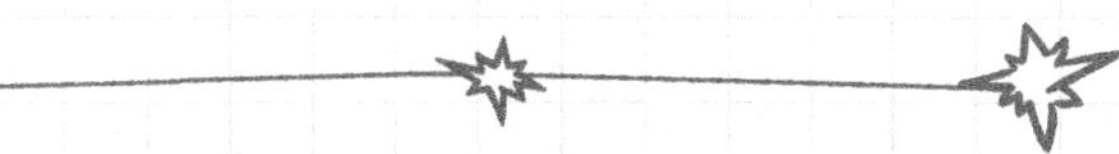

4. Miles collects bugs. He has a worm that measures $7\frac{3}{10}$ cm and a caterpillar that measures $4\frac{6}{10}$ cm. How much longer is the worm than the caterpillar?

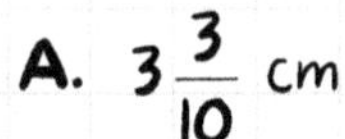

A. $3\frac{3}{10}$ cm

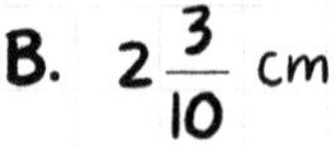

B. $2\frac{3}{10}$ cm

C. $3\frac{7}{10}$ cm

D. $2\frac{7}{10}$ cm

SHOW YOUR WORK

5. Sara's garden has many flowers. One flower in her garden has a stem that measures $12\frac{3}{10}$ cm and a flower that measures $6\frac{4}{10}$ cm. How tall is the flower total?

A. $18\frac{7}{10}$

B. $6\frac{1}{10}$

C. $16\frac{1}{10}$

D. $18\frac{3}{10}$

SHOW YOUR WORK

NOTES

Do you know there are other ways to represent fractions? We can use multiplication properties to represent fractions different ways. For example, the number 5 is the same as the number 5 x 1. We can do the same thing with fractions. If I have the number $\frac{7}{8}$, we can take the base unit of $\frac{1}{8}$ and multiple it by 7 because $\frac{7}{8}$ is 7 of the base units.

Let's look at another example:

$\frac{5}{6}$

What is the base unit? $\frac{1}{6}$

How many base units are there? 5

So $\frac{5}{6}$ is the same thing as $5 \times \frac{1}{6}$.

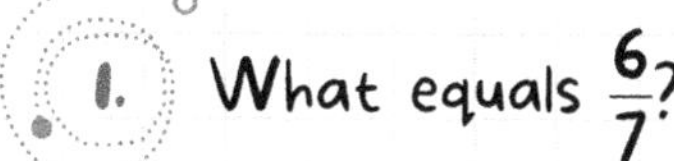

1. What equals $\frac{6}{7}$?

A. $6 \times \frac{1}{7}$

B. $67 \times \frac{1}{6}$

C. $7 \times \frac{1}{6}$

D. $\frac{1}{6} \times \frac{1}{7}$

SHOW YOUR WORK

2. What equals $\frac{3}{4}$?

A. $4 \times \frac{1}{4}$

B. $3 \times \frac{1}{3}$

C. $4 \times \frac{1}{3}$

D. $3 \times \frac{1}{4}$

3. What is the base unit of $\frac{8}{12}$?

A. $\frac{1}{8}$

B. $\frac{2}{6}$

C. $\frac{1}{12}$

D. $\frac{2}{12}$

4. What is the base unit of $\frac{13}{20}$?

A. $\frac{1}{13}$

B. $\frac{1}{20}$

C. $\frac{20}{13}$

D. $\frac{2}{13}$

SHOW YOUR WORK

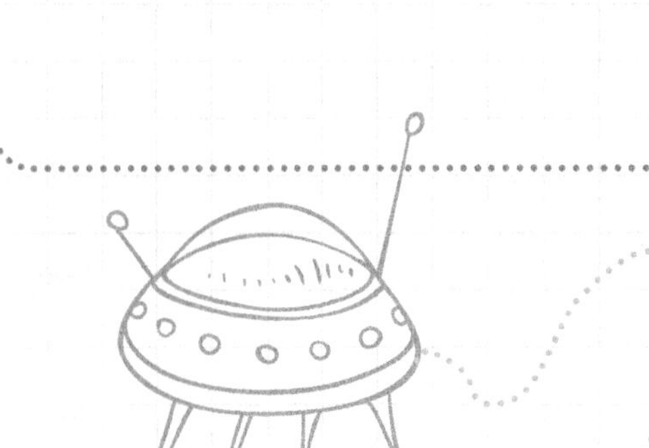

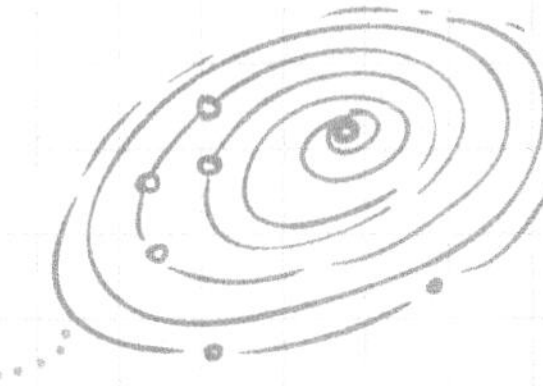

5. What is $7 \times \frac{1}{10}$?

A. $\frac{7}{10}$

B. $\frac{10}{7}$

C. $\frac{70}{10}$

D. $\frac{10}{70}$

NOTES

ARGOPREP

3.2.F | Build fractions from unit fractions

Can we extend the information we learned before to begin multiplying fractions? Yes! If when we take a base unit by a number, we get that number of the base unit, we can assume that when we multiply a fraction by a whole number, we will only manipulate the numerator of the fraction, the denominator stays the same. **We can always rewrite the problem as multiplication of a whole number by a base unit.**

Let's review by looking at some real numbers.

$5 \times \frac{3}{4} = ?$

First, we can rewrite the problem as $15 \times \frac{1}{4}$.

The answer would be $\frac{15}{4}$.

When you are solving these problems, follow the steps below.

First, take the whole number times the numerator.

Next, multiply the answer by the base unit to express as a fraction.

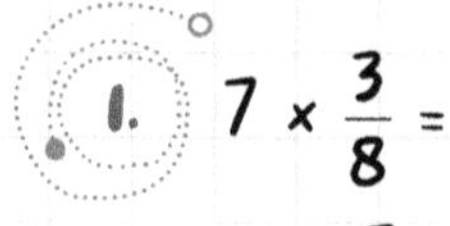

1. $7 \times \frac{3}{8} =$

A. $\frac{7}{24}$

B. $\frac{8}{21}$

C. $\frac{21}{56}$

D. $\frac{21}{8}$

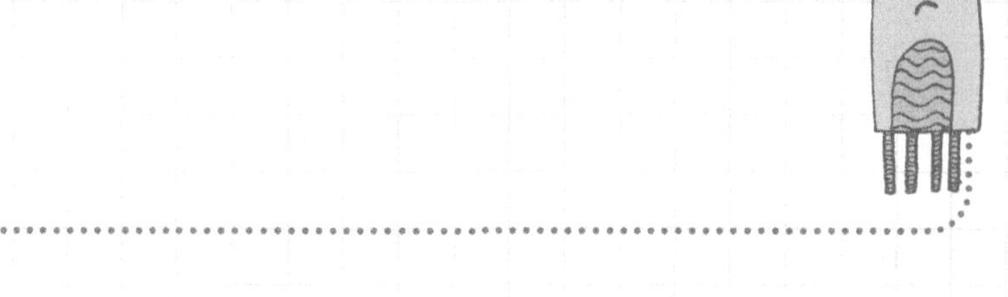

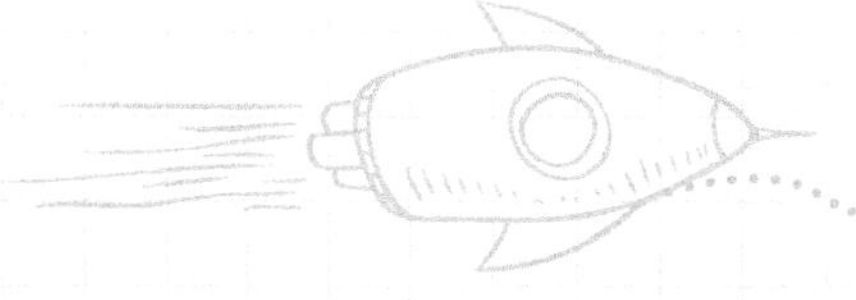

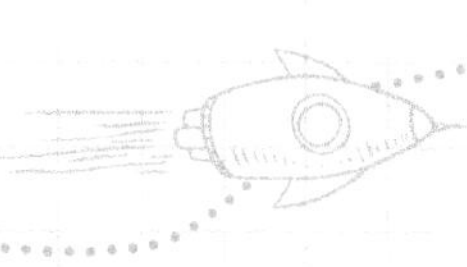

2. $2 \times \frac{5}{6} =$

A. $\frac{12}{10}$

B. $\frac{10}{12}$

C. $\frac{10}{6}$

D. $\frac{6}{10}$

SHOW YOUR WORK

3. $4 \times \frac{11}{12} =$

A. $\frac{12}{44}$

B. $\frac{44}{12}$

C. $\frac{44}{48}$

D. $\frac{45}{12}$

SHOW YOUR WORK

4. $3 \times \frac{2}{9} =$

A. $\frac{6}{9}$

B. $\frac{9}{6}$

C. $\frac{6}{27}$

D. $\frac{27}{6}$

SHOW YOUR WORK

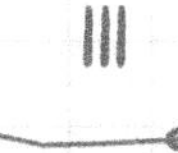

5. $6 \times \frac{2}{7} =$

A. $\frac{8}{7}$

B. $\frac{7}{12}$

C. $\frac{12}{42}$

D. $\frac{12}{7}$

SHOW YOUR WORK

NOTES

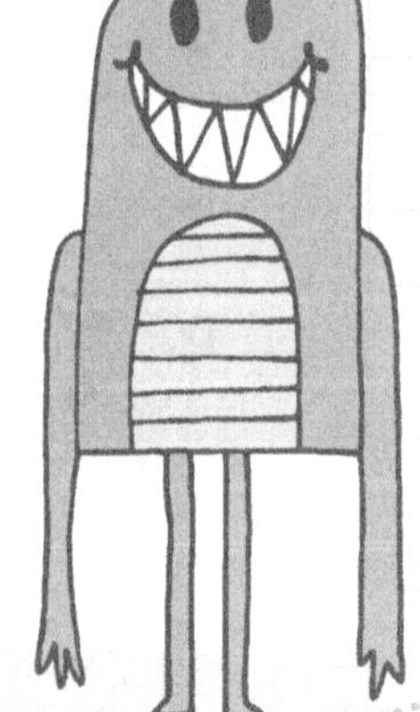

Do you know how to apply these skills in word problems? Remember the following steps when you are trying to solve word problems:

1 → Read the word problem.

2 → Reread the problem and circle what you want to know.

3 → Reread the problem and underline what you know.

4 → Write a problem based on what you know.

5 → Apply the skills you have learned so far to solve the problem and check your answer.

Remember to phrase your answer in terms of the real world.

1. Maria makes a can of soup for lunch. Each can holds $\frac{3}{4}$ cup of soup. How much soup has she eaten after 4 days?

A. 3 cups

B. $\frac{12}{4}$ cups

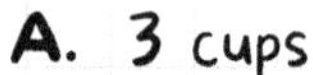

C. $\frac{12}{16}$ cups

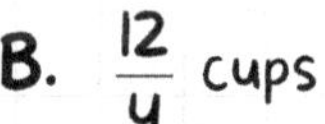

D. $\frac{4}{12}$ cups

SHOW YOUR WORK

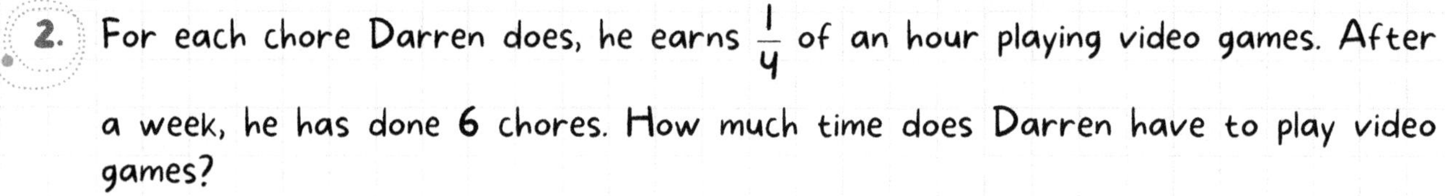

2. For each chore Darren does, he earns $\frac{1}{4}$ of an hour playing video games. After a week, he has done **6** chores. How much time does Darren have to play video games?

A. $\frac{24}{6}$ hour

B. $\frac{4}{6}$ hour

C. $\frac{6}{24}$ hour

D. $\frac{6}{4}$ hour

SHOW YOUR WORK

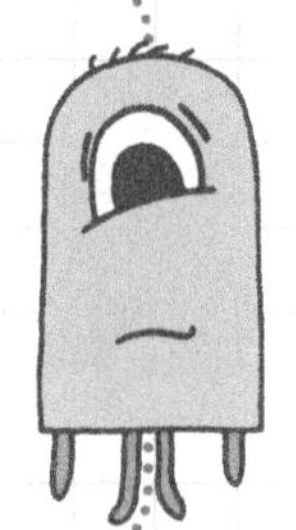

3. Evie ran $\frac{3}{5}$ of a mile yesterday. Today, she ran two times as far. How far did she run today?

A. $\frac{6}{5}$ miles

B. $\frac{5}{6}$ miles

C. $\frac{9}{5}$ miles

D. $\frac{9}{25}$ miles

SHOW YOUR WORK

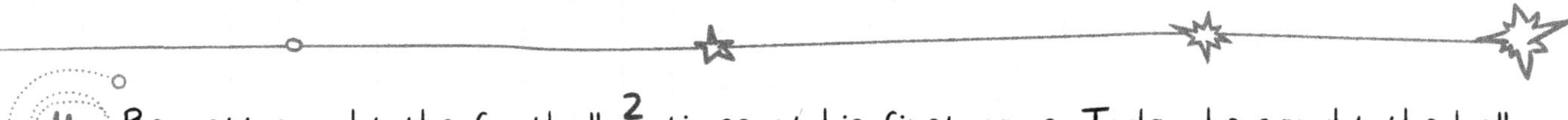

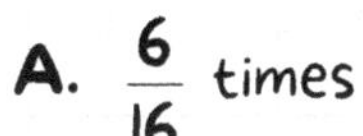

4. Bennett caught the football $\frac{2}{14}$ times at his first game. Today, he caught the ball three times as much. How many catches did he make today?

A. $\frac{6}{16}$ times

B. $\frac{4}{14}$ times

C. $\frac{6}{14}$ times

D. $\frac{14}{6}$ times

5. Four friends are at an amusement park. Each friend has $\frac{1}{2}$ of the money needed to buy a book of tickets. How many books of tickets can they buy?

A. 3 books

B. 1 book

C. 2 books

D. 4 books

SHOW YOUR WORK

3.3.A | Understand decimal notation for fractions, and compare decimal fractions

Let's look at some equivalent fractions. **Fractions with a denominator 10, are equal to fractions with the denominator 100 as long as you multiply the numerator by 10.** For example $\frac{4}{10}$ is the same as $\frac{4}{100}$. We can use this information when we are trying to add fractions with denominators of 10 and 100.

Let's review. $\frac{2}{10} + \frac{6}{100} = ?$

$\frac{2}{10}$ is the same as $\frac{20}{100}$ and now we can combine both fractions because they have equal denominators.

$\frac{20}{100} + \frac{6}{100} = \frac{26}{100}$ SO $\frac{2}{10} + \frac{6}{700} = \frac{26}{100}$

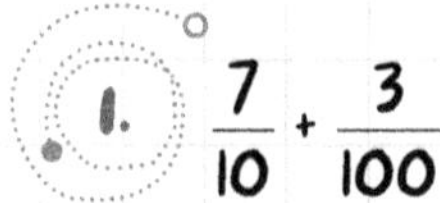

1. $\frac{7}{10} + \frac{3}{100} =$

A. $\frac{10}{100}$

B. $\frac{73}{100}$

C. $\frac{37}{100}$

D. $\frac{21}{100}$

2. $\frac{2}{10} + \frac{62}{100} =$

A. $\frac{62}{110}$

B. $\frac{60}{100}$

C. $\frac{64}{100}$

D. $\frac{82}{100}$

3. $\frac{3}{10} + \frac{6}{100} =$

A. $\frac{36}{100}$

B. $\frac{63}{100}$

C. $\frac{9}{100}$

D. $\frac{63}{1000}$

SHOW YOUR WORK

4. $\frac{9}{10} + \frac{1}{100} =$

A. $\frac{8}{100}$

B. $\frac{10}{100}$

C. $\frac{91}{100}$

D. $\frac{19}{100}$

5. $\frac{4}{10} + \frac{24}{100} =$

A. $\frac{64}{100}$

B. $\frac{28}{100}$

C. $\frac{20}{100}$

D. $\frac{28}{110}$

SHOW YOUR WORK

6. $\frac{1}{10} + \frac{7}{100} =$

A. $\frac{8}{100}$

B. $\frac{17}{100}$

C. $\frac{71}{100}$

D. $\frac{8}{10}$

SHOW YOUR WORK

7. $\frac{5}{10} + \frac{18}{100} =$

A. $\frac{28}{100}$

B. $\frac{13}{100}$

C. $\frac{23}{100}$

D. $\frac{68}{100}$

SHOW YOUR WORK

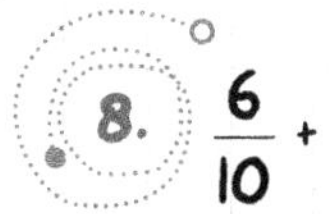

8. $\frac{6}{10} + \frac{33}{100} =$

A. $\frac{27}{100}$

B. $\frac{39}{100}$

C. $\frac{93}{100}$

D. $\frac{36}{100}$

SHOW YOUR WORK

9. $\frac{8}{10} + \frac{4}{100} =$

A. $\frac{84}{100}$

B. $\frac{12}{100}$

C. $\frac{48}{100}$

D. $\frac{76}{100}$

SHOW YOUR WORK

10. $\frac{9}{10} + \frac{2}{100} =$

A. $\frac{90}{100}$

B. $\frac{12}{100}$

C. $\frac{11}{100}$

D. $\frac{92}{100}$

SHOW YOUR WORK

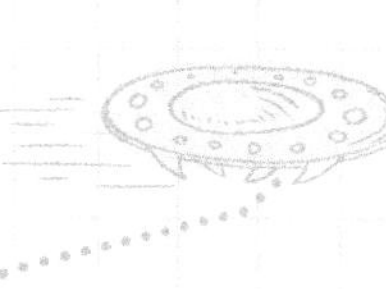

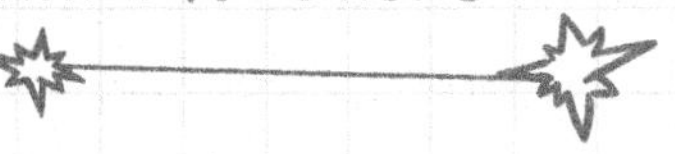

We can use our knowledge of fractions to express numbers as decimals. **Fractions with a denominator of 10 are equal to decimals to the first place right of the decimal.**

For example $\frac{4}{10} = 0.4$

Fractions with a denominator of 100 are equal to decimals to the second place right of the decimal.

For example $\frac{78}{100} = 0.78$

1. Write the equivalent fraction to 0.54.

A. $\frac{5}{100}$

B. $\frac{4}{10}$

C. $\frac{5}{10}$

D. $\frac{54}{100}$

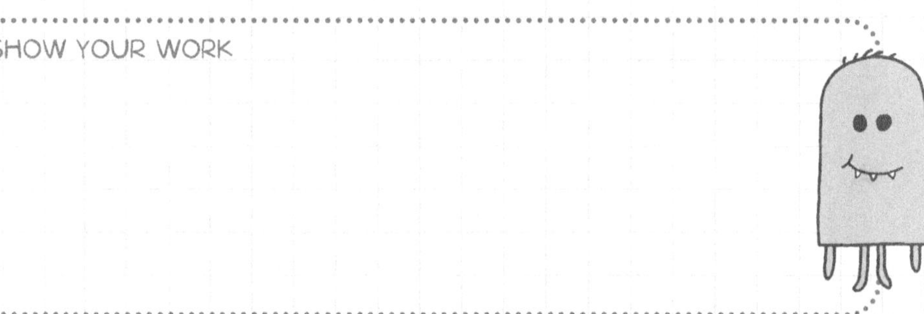

2. Write the equivalent fraction to 0.2.

A. $\frac{2}{10}$

B. $\frac{1}{10}$

C. $\frac{2}{100}$

D. $\frac{1}{100}$

SHOW YOUR WORK

3. Write the equivalent fraction to 0.98.

A. $\frac{9}{10}$

B. $\frac{98}{100}$

C. $\frac{8}{100}$

D. $\frac{8}{10}$

SHOW YOUR WORK

4. Write the equivalent fraction to 0.17.

A. $\frac{10}{100}$

B. $\frac{7}{100}$

C. $\frac{17}{100}$

D. $\frac{10}{100}$

SHOW YOUR WORK

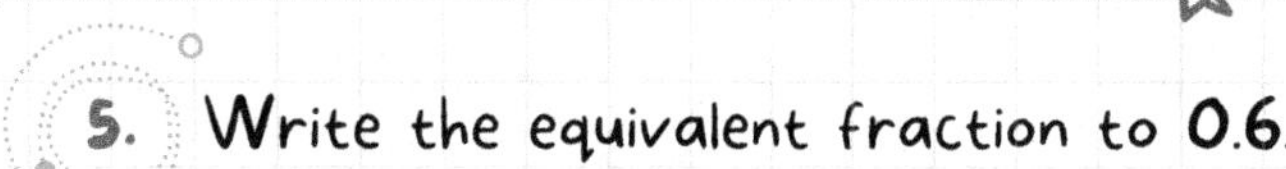

5. Write the equivalent fraction to 0.6.

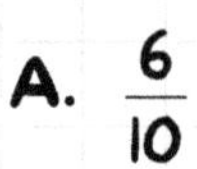

A. $\frac{6}{10}$

B. $\frac{6}{100}$

C. $\frac{60}{100}$

D. $\frac{60}{1000}$

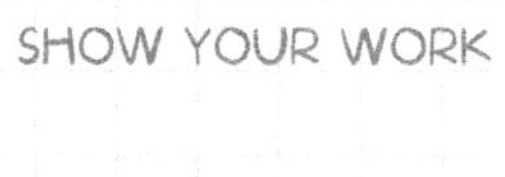

SHOW YOUR WORK

6. Write the equivalent decimal to $\frac{2}{10}$.

A. 0.2

B. 0.02

C. 0.21

D. 0.12

SHOW YOUR WORK

7. Write the equivalent decimal to $\frac{72}{100}$.

A. 0.2

B. 0.7

C. 0.27

D. 0.72

SHOW YOUR WORK

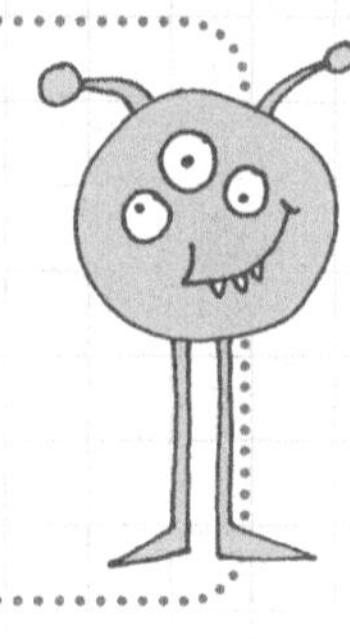

8. Write the equivalent decimal to $\frac{8}{10}$.

A. 0.1

B. 0.08

C. 0.8

D. 0.18

SHOW YOUR WORK

9. Write the equivalent decimal to $\frac{36}{100}$.

A. 0.36

B. 0.63

C. 0.3

D. 0.6

SHOW YOUR WORK

10. Write the equivalent decimal to $\frac{97}{100}$.

A. 0.9

B. 0.97

C. 0.7

D. 0.79

SHOW YOUR WORK

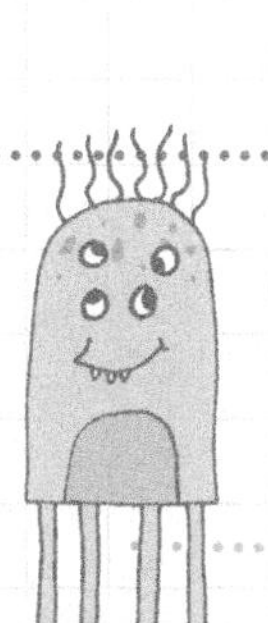

3.3.C Understand decimal notation for fractions, and compare decimal fractions

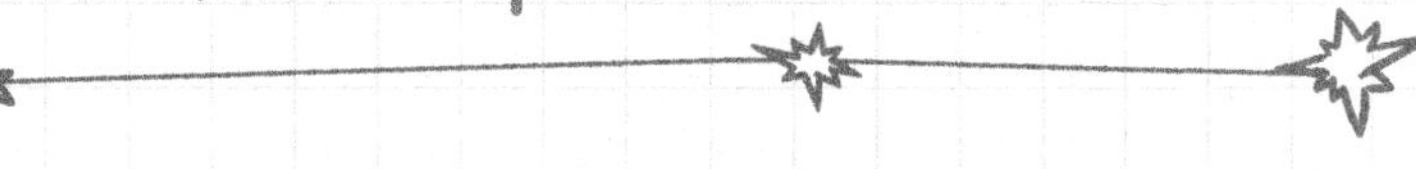

We can use the information we have learned about decimals to compare them to each other. Which is larger $\frac{1}{10}$ or $\frac{1}{100}$? $\frac{1}{10}$ is larger because the whole is divided into fewer pieces. If $\frac{1}{10}$ is larger and 0.1 is equal to $\frac{1}{10}$ then 0.1 is larger than 0.01. **When comparing numbers, first you should consider the value of the tenths place and then look at the hundredths place.**

Let's review.
Which number is larger 0.65 or 0.8?

Remember, first review the tenths place. Is 8 larger than 6? Yes! So 0.8 is larger than 0.65.

Remember our signs when comparing two decimals.

$<$: Less than (means the number on the left is smaller than the number on the right)

$>$: Greater than (means the number on the left is larger than the number on the right)

$\leq$: Less than or equal to (means the number on the left is smaller than or equal to the number on the right)

$\geq$: Greater than or equal to (means the number on the left is larger than or equal to the number on the right)

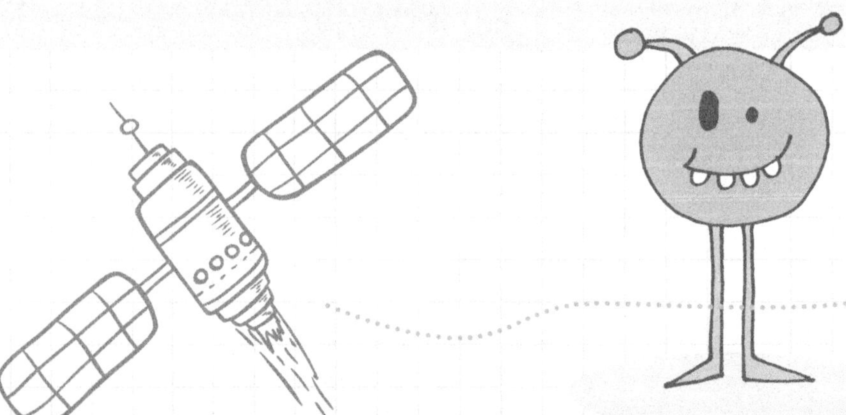

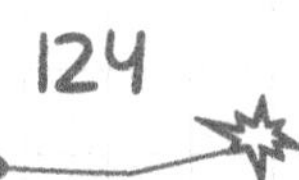

3.3.C | Understand decimal notation for fractions, and compare decimal fractions

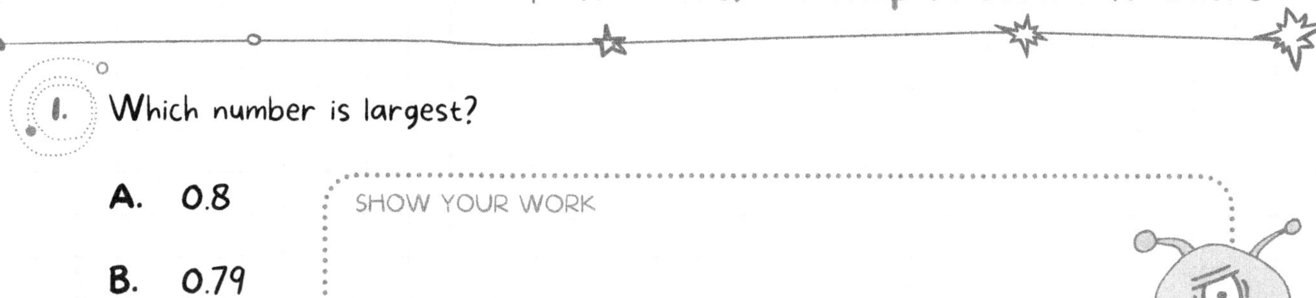

1. Which number is largest?

A. 0.8

B. 0.79

C. 0.65

D. 0.18

SHOW YOUR WORK

2. Which number is smallest?

A. 0.2

B. 0.8

C. 0.35

D. 0.17

SHOW YOUR WORK

3. Which number is largest?

A. 0.53

B. 0.5

C. 0.67

D. 0.32

SHOW YOUR WORK

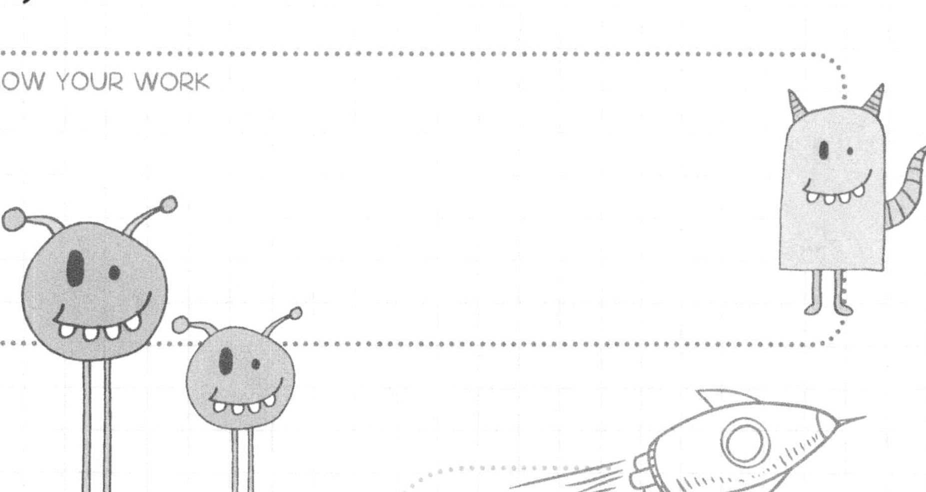

4. Which number is smallest?

A. 0.4

B. 0.6

C. 0.54

D. 0.8

SHOW YOUR WORK

5. Which number is largest?

A. 0.49

B. 0.79

C. 0.7

D. 0.65

SHOW YOUR WORK

6. 0.65 ______ 0.36

A. $>$

B. $<$

C. $=$

D. $\geq$

SHOW YOUR WORK

7. 0.5 _____ 0.74

A. >

B. <

C. =

D. ≥

SHOW YOUR WORK

8. 0.42 _____ 0.8

A. >

B. <

C. =

D. ≥

SHOW YOUR WORK

9. 0.63 _____ 0.23

A. >

B. <

C. =

D. ≥

SHOW YOUR WORK

10. 0.9 ______ 0.45

A. >

B. <

C. =

D. ≥

SHOW YOUR WORK

NOTES

3.4. Chapter Test

1. Which number could replace n to provide an equivalent fraction?

$\frac{n}{12} = \frac{6}{36}$

A. 1

B. 2

C. 3

D. 4

SHOW YOUR WORK

2. When comparing $\frac{3}{6}$ and $\frac{3}{18}$, which common denominator should you use?

A. 6

B. 12

C. 18

D. 24

SHOW YOUR WORK

3. $\frac{4}{5}$ ------ $\frac{7}{8}$

A. $>$

B. $<$

C. $=$

D. $\geq$

SHOW YOUR WORK

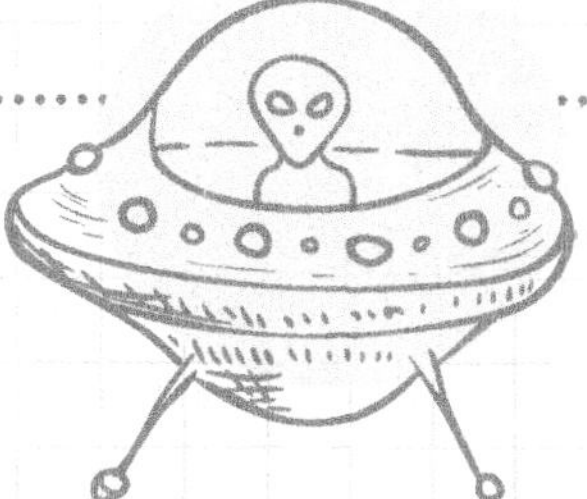

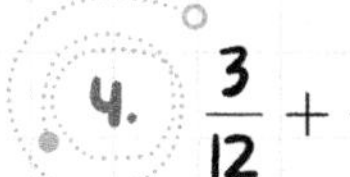

4. $\frac{3}{12} + \frac{5}{12}$

A. $\frac{8}{12}$

B. $\frac{8}{24}$

C. $\frac{15}{24}$

D. $\frac{2}{12}$

SHOW YOUR WORK

5. A pizza was divided into 12 pieces. My mom and dad each ate 2 pieces and each of my two sisters ate a piece. What problem represents how many pieces were left for me?

SHOW YOUR WORK

6. Which fraction problem is equivalent to $\frac{12}{18}$?

A. $\frac{1}{18} + \frac{1}{12}$

B. $\frac{12}{18} + \frac{18}{12}$

C. $\frac{6}{18} + \frac{6}{12}$

D. $\frac{6}{18} + \frac{6}{18}$

SHOW YOUR WORK

7. Which fraction problem is not equivalent to $2\frac{1}{3}$?

A. $\frac{3}{3} + \frac{3}{3} + \frac{1}{3}$

B. $\frac{1}{3} + \frac{3}{3}$

C. $\frac{2}{3} + \frac{1}{3} + \frac{2}{3} + \frac{2}{3}$

D. $\frac{1}{3} + \frac{1}{3} + \frac{1}{3} + \frac{1}{3} + \frac{3}{3}$

SHOW YOUR WORK

8. $6\frac{3}{7} + 3\frac{2}{7} =$

A. $3\frac{5}{7}$

B. $3\frac{1}{7}$

C. $9\frac{5}{7}$

D. $9\frac{4}{7}$

SHOW YOUR WORK

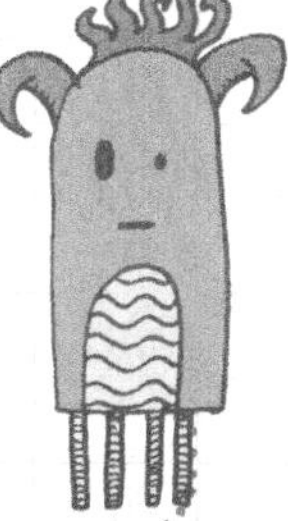

9. $7\frac{1}{8} - 3\frac{3}{8} =$

A. $4\frac{2}{8}$

B. $3\frac{2}{8}$

C. $3\frac{6}{8}$

D. $4\frac{6}{8}$

SHOW YOUR WORK

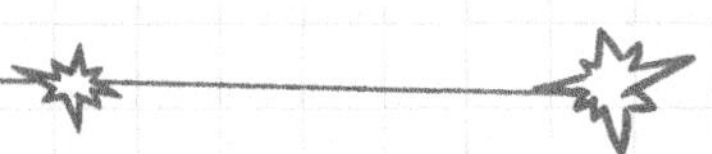

10. Morgan makes a plan for his afterschool time. He wants to spend $\frac{1}{2}$ hour working on homework and $\frac{1}{2}$ hour doing his chores. He also exercises for $\frac{1}{2}$ hour and practices his piano for $\frac{1}{2}$ hour. If he has 3 hours before dinner, how much time does he have left to play outside?

A. 1 hour

B. 2 hours

C. $1\frac{1}{2}$ hour

D. $2\frac{1}{2}$ hour

SHOW YOUR WORK

11. What is the base unit of $\frac{8}{15}$?

A. $\frac{1}{8}$

B. $\frac{1}{15}$

C. $\frac{2}{8}$

D. $\frac{3}{15}$

SHOW YOUR WORK

12. What is $9 \times \frac{1}{20}$?

A. $\frac{20}{9}$

B. $\frac{9}{2}$

C. $\frac{9}{20}$

D. $\frac{7}{13}$

SHOW YOUR WORK

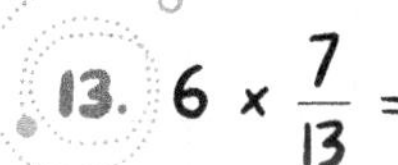

13. $6 \times \frac{7}{13} =$

A. $\frac{1}{13}$

B. $\frac{6}{13}$

C. $\frac{42}{78}$

D. $\frac{42}{13}$

SHOW YOUR WORK

14. Graham is painting a forest scene. He uses $\frac{3}{5}$ of a tube of green paint to paint a tree. He wants to paint 10 trees. How many tubes of green paint does he need to buy?

A. 6 tubes

B. 5 tubes

C. 4 tubes

D. 3 tubes

SHOW YOUR WORK

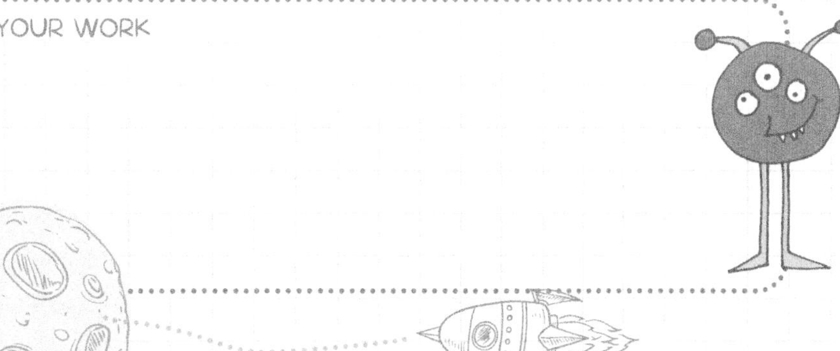

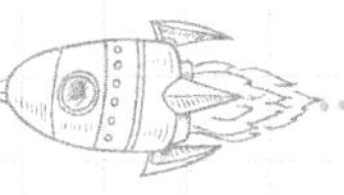

15. $\frac{2}{10} + \frac{7}{100} =$

A. $\frac{9}{100}$

B. $\frac{72}{100}$

C. $\frac{27}{100}$

D. $\frac{90}{100}$

SHOW YOUR WORK

16. $\frac{8}{10} + \frac{15}{100} =$

A. $\frac{95}{100}$

B. $\frac{23}{100}$

C. $\frac{20}{100}$

D. $\frac{16}{100}$

SHOW YOUR WORK

17. Write the equivalent fraction to 0.32.

A. $\frac{10}{32}$

B. $\frac{1}{32}$

C. $\frac{32}{10}$

D. $\frac{32}{100}$

SHOW YOUR WORK

3.4. Chapter Test

18. Write the equivalent decimal to $\frac{5}{10}$.

A. 0.1

B. 0.15

C. 0.5

D. 0.51

SHOW YOUR WORK

19. Which number is smallest?

A. 0.2

B. 0.18

C. 0.6

D. 0.26

SHOW YOUR WORK

20. 0.75 ______ 0.3

A. >

B. <

C. =

D. ≥

SHOW YOUR WORK

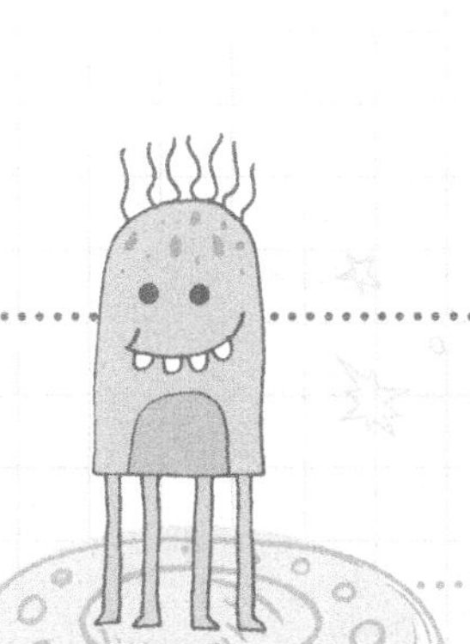

Chapter 4 : Measurement & Data

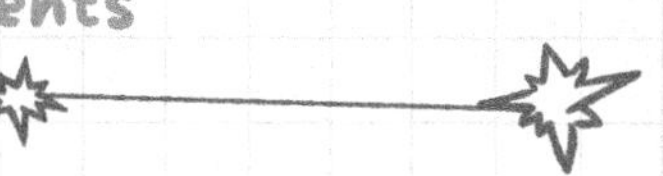

When we consider measurements, we need to know conversion factors for different measurements. **Conversions let us change a measurement from one unit to another unit.**

100 cm = 1 m

1000 m = 1 km

1000 g = 1 kg

16 oz = 1 pound

1000 ml = 1 L

60 sec = 1 min

60 min = 1 hour

We can use our math skills to apply conversions to real world examples.

1. Which measurement is equal to 40 minutes?

A. 240 hours

B. 24 hours

C. 2400 seconds

D. 24 seconds

SHOW YOUR WORK

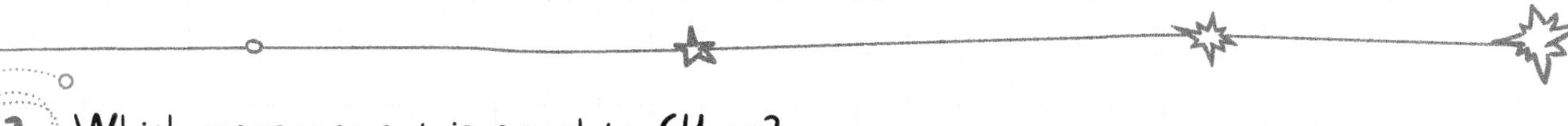

2. Which measurement is equal to 64 oz?

A. 4 grams

B. 4 pounds

C. 4 liters

D. 4 meters

SHOW YOUR WORK

3. Which measurement is equal to 7 minutes?

A. 420 seconds

B. 420 hours

C. 4.2 seconds

D. 4.2 hours

SHOW YOUR WORK

4. Which measurement is equal to 8000 grams?

A. 0.8 kilograms

B. 800 kilograms

C. 80 kilograms

D. 8 kilograms

SHOW YOUR WORK

5. Which measurement is equal to 3 liters?

A. 30 milliliters

B. 3 milliliters

C. 300 milliliters

D. 3000 milliliters

SHOW YOUR WORK

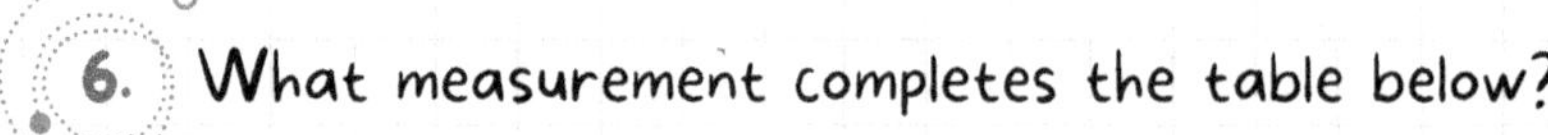

6. What measurement completes the table below?

seconds	**minutes**
7200	

A. 12

B. 1.2

C. 120

D. 1200

SHOW YOUR WORK

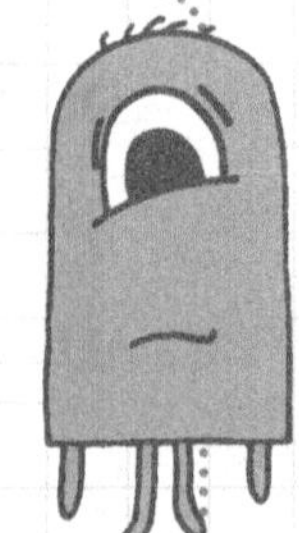

7. What measurement completes the table below?

milliliters	liters
8500	

A. 85

B. 8.5

C. 850

D. 0.85

SHOW YOUR WORK

8. What measurement completes the table below?

grams	kilograms
6700	

A. 67

B. 6.7

C. 670

D. 0.67

SHOW YOUR WORK

9. What measurement completes the table below?

centimeters	meters
	7.4

A. 740

B. 74

C. 7400

D. 700

SHOW YOUR WORK

10. What measurement completes the table below?

ounces	pounds
	8

A. 1280

B. 1.28

C. 12.8

D. 128

SHOW YOUR WORK

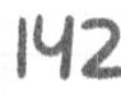

Do you know how to apply these skills in word problems? Remember the following steps when you are trying to solve word problems:

1 → Read the word problem.

2 → Reread the problem and circle what you want to know.

3 → Reread the problem and underline what you know.

4 → Write a problem based on what you know.

5 → Apply the skills you have learned so far to solve the problem and check your answer.

Remember to phrase your answer in terms of the real world.

1. What is the first step you should complete to solve the below problem?

Mary makes 3 gallons of iced tea. She has some friends over and pours some iced tea into a 2 quart container. If she and her friends drink all the iced tea, how much tea is left? (Remember, 4 quarts = 1 gallon).

A. Convert iced tea gallons to quarts.

B. Subtract 2 quarts from 3 gallons.

C. Add 2 quarts to 3 gallons.

D. Convert iced tea quarts to gallons.

2. Mary makes 3 gallons of iced tea. She has some friends over and pours some iced tea into a 2 quart container. If she and her friends drink all the iced tea, how much tea is left? (Remember, 4 quarts = 1 gallon).

A. 10 quarts

B. 10 gallons

C. 14 quarts

D. 14 gallons

SHOW YOUR WORK

3. One pumpkin weighs 8 pounds 3 ounces. A second pumpkin weighs 4 pounds 13 ounces. A third pumpkin weighs 6 pounds 8 ounces. How much do all the pumpkins weight?

A. 18 pounds

B. 18.5 pounds

C. 19 pounds

D. 19.5 pounds

SHOW YOUR WORK

4. Ryan spent 45 minutes on his math homework, 1 hour and 20 minutes on his book report and 25 minutes studying his spelling words. How long did Ryan spend studying?

A. 2 hours

B. 2 hours 15 minutes

C. 2 hours 30 minutes

D. 2 hours 45 minutes

SHOW YOUR WORK

5. On Monday, our family walked 2.1 kilometers. On Friday, we walked twice as long as Monday. On Sunday, we walked 3.2 kilometers. How far did we walk Monday-Sunday?

A. 5.3 kilometers

B. 6.6 kilometers

C. 7.6 meters

D. 9.5 kilometers

SHOW YOUR WORK

6. Miguel is tired of carrying a heavy backpack. He wants to prove it is too heavy. One school day, he is carrying the backpack (1 lb 15 oz), his math book (3 lb 5 oz), his laptop (6 lb 2 oz) and his pencil case (14 oz). How much weight is he carrying?

A. 12 lbs 4 oz

B. 11 lbs 15 oz

C. 12 lbs 2 oz

D. 11 lbs 4 oz

SHOW YOUR WORK

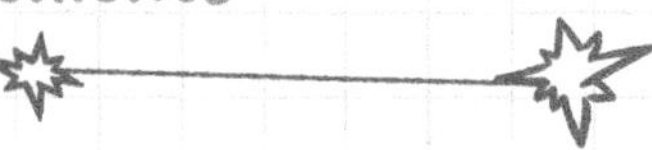

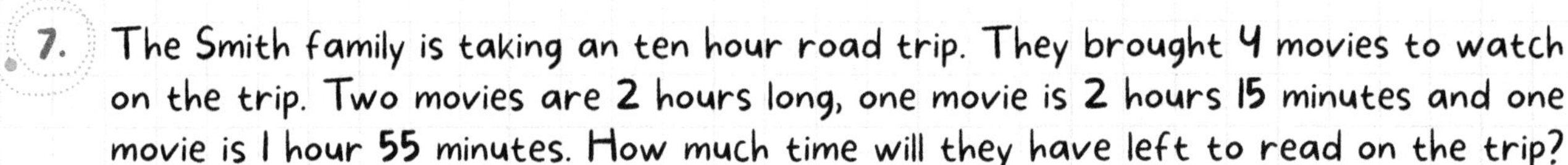

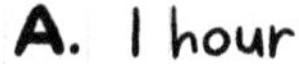

7. The Smith family is taking an ten hour road trip. They brought 4 movies to watch on the trip. Two movies are 2 hours long, one movie is 2 hours 15 minutes and one movie is 1 hour 55 minutes. How much time will they have left to read on the trip?

A. 1 hour

B. 1 hour 50 mins

C. 8 hours 10 mins

D. 8 hours

SHOW YOUR WORK

8. The length of Eric's bedroom is 12 feet, 9 inches. The width of the room is 2 feet, 8 inches less than the length. What is the width of Eric's bedroom?

A. 10 feet, 1 inch

B. 10 feet

C. 10 feet, 5 inches

D. 10 feet, 2 inches

SHOW YOUR WORK

9. Emily and Christopher are running a relay race. Christopher ran his leg in 32 minutes. Emily took 9 minutes longer to run her leg. How long did it take them to run the race in all?

A. 41 minutes

B. 50 minutes

C. 1 hour 13 minutes

D. 1 hour 3 minutes

SHOW YOUR WORK

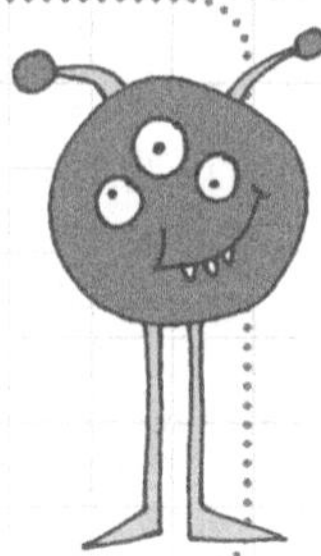

10. Amanda spent 45 minutes exercising last week. Missy spent double the amount of time exercising. How much time did Missy spend exercising?

A. 1 hour

B. 1 hour 30 minutes

C. 2 hours

D. 45 minutes

SHOW YOUR WORK

NOTES

It is possible to use our knowledge of perimeter and area to solve problems using information we know. Let's look at an example.

First, remember the perimeter of a rectangle is 2xlength + 2x width. The area of a rectangle is length x width.

We can use these formulas and what we know, to find what we don't know.

For example, the perimeter of a garden is measured as 44 ft. The length of the garden is 16 feet. How long is the width?

First, if the length of the garden is 16 feet, 2 x16 = 34.

The perimeter is 44 - 34, leaves 10 feet for double the width.

The width would be 5 feet.

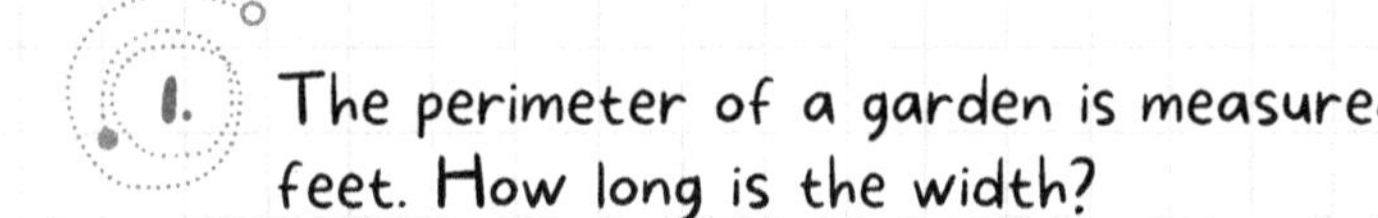

1. The perimeter of a garden is measured as 56 ft. The length of the garden is 20 feet. How long is the width?

A. 4 ft

B. 5 ft

C. 6 ft

D. 8 ft

SHOW YOUR WORK

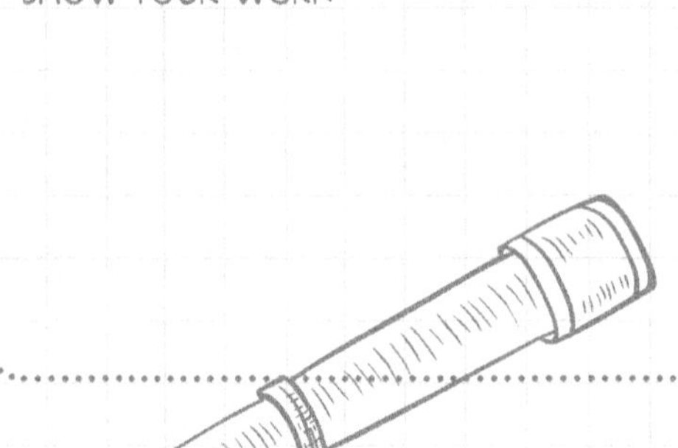

2. The perimeter of a garden is measured as 24 ft. The width of the garden is 6 feet. How long is the length?

A. 12 ft

B. 10 ft

C. 4 ft

D. 6 ft

SHOW YOUR WORK

3. The area of a garden is measured as 80 ft^2. The length of the garden is 8 feet. How long is the width?

A. 10 ft

B. 12 ft

C. 14 ft

D. 16 ft

SHOW YOUR WORK

4. The area of a garden is measured as 44 ft^2. The length of the garden is 11 feet. How long is the width?

A. 2 ft

B. 4 ft

C. 6 ft

D. 8 ft

SHOW YOUR WORK

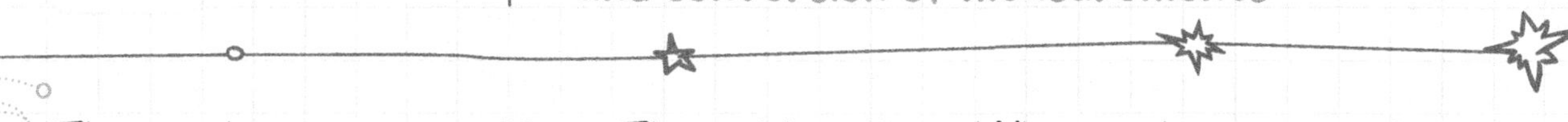

5. The length of a room is 10 ft. The width is 12 ft. What is the perimeter?

A. 62 ft

B. 44 ft

C. 120 ft

D. 22 ft

SHOW YOUR WORK

6. The length of a room is 8 ft. The width is 6 ft. What is the area?

A. 42 ft^2

B. 28 ft^2

C. 48 ft^2

D. 26 ft^2

SHOW YOUR WORK

7. If a rectangle has a length of 6 and an area of 72, what is its width?

A. 12

B. 14

C. 16

D. 18

SHOW YOUR WORK

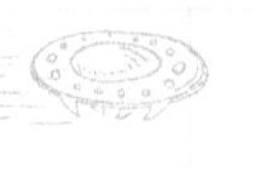

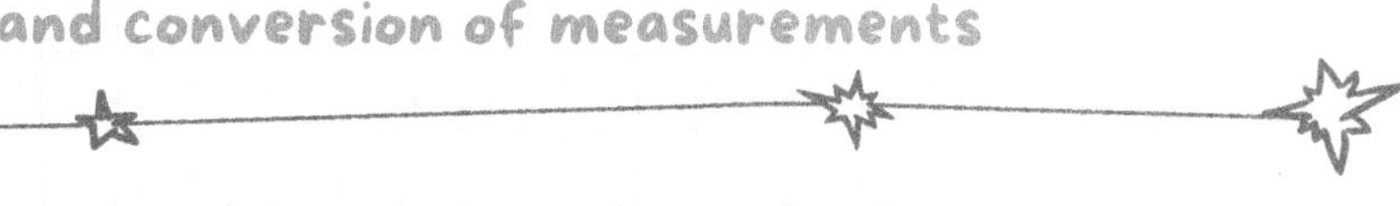

8. If a rectangle has a width of 4 and a perimeter of 18, what is its length?

A. 5

B. 6

C. 7

D. 8

SHOW YOUR WORK

9. If a rectangle has a length of 8 and an area of 64, what is its perimeter?

A. 32

B. 34

C. 8

D. 10

SHOW YOUR WORK

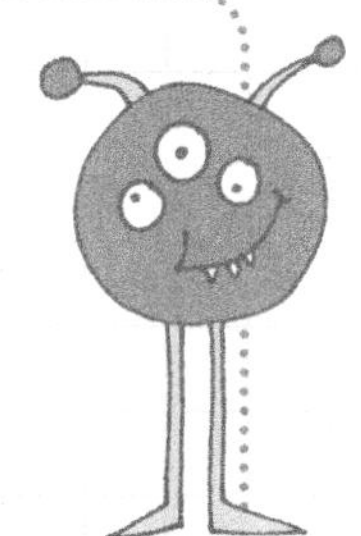

10. If a rectangle has a width of 12 and a perimeter of 44, what is the area?

A. 100

B. 120

C. 130

D. 145

SHOW YOUR WORK

4.2. Represent and interpret data

Some times when we are given a variety of measurements, a line plot is helpful to represent different amounts of information in a visual way. A line plot can show where data falls.

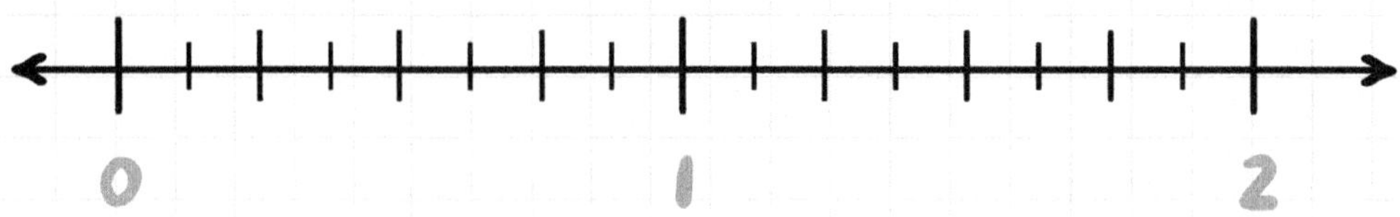

We can use line plots to analyze data. Each whole number can be divided into equi-distant marks that mark $\frac{1}{2}$, $\frac{1}{4}$ and $\frac{1}{8}$.

1. Using the line plot below, where would you graph $4\frac{1}{4}$?

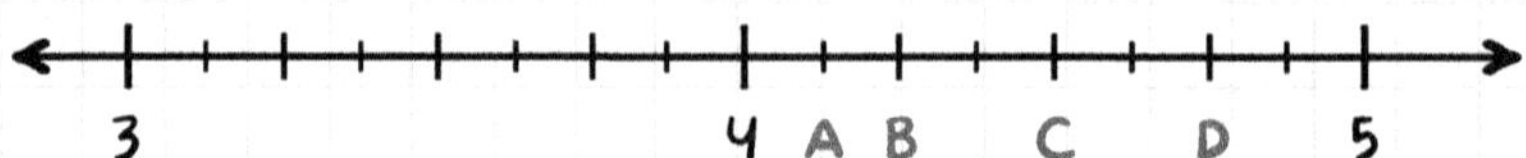

A. Point A

B. Point B

C. Point C

D. Point D

SHOW YOUR WORK

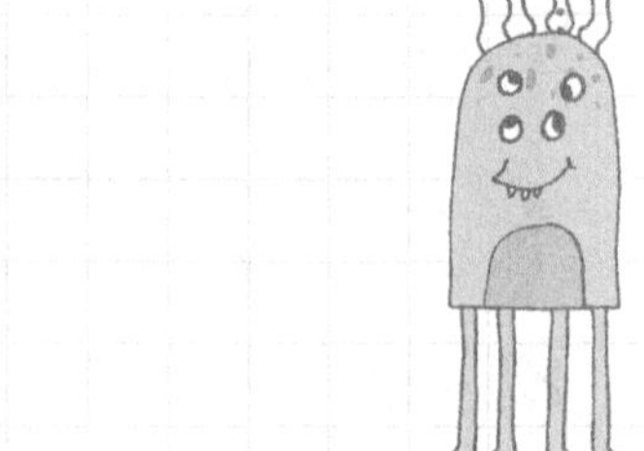

2. Using the line plot below, where would you graph $10\frac{4}{8}$?

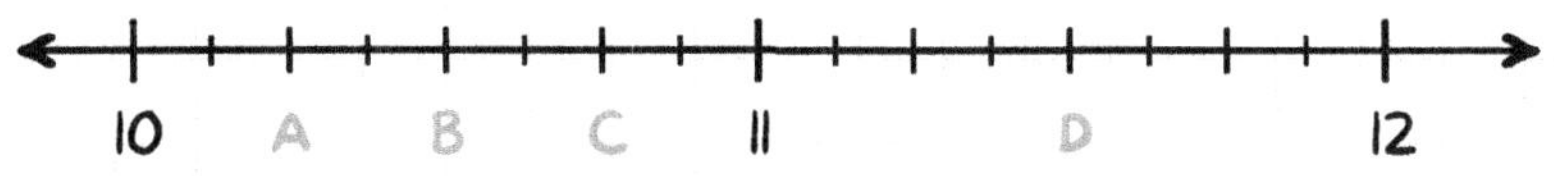

A. Point A

B. Point B

C. Point C

D. Point D

SHOW YOUR WORK

3. Using the line plot below, where would you graph $9\frac{3}{4}$?

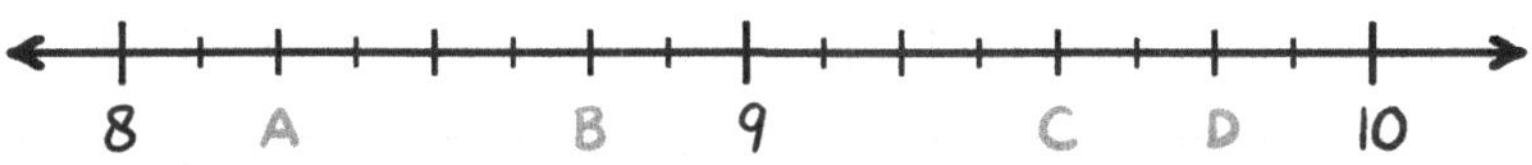

A. Point A

B. Point B

C. Point C

D. Point D

SHOW YOUR WORK

4. Using the line plot below, where would you graph $20\frac{1}{8}$?

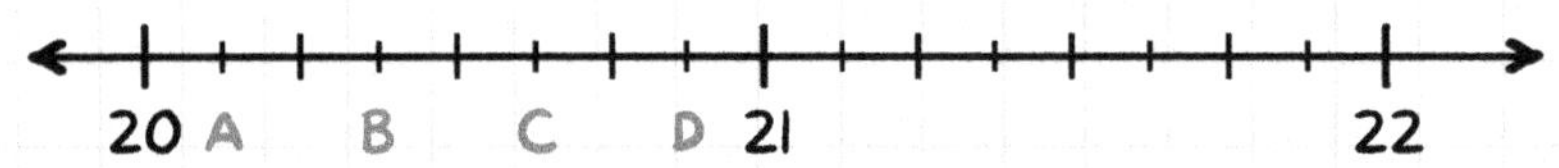

A. Point A

B. Point B

C. Point C

D. Point D

SHOW YOUR WORK

5. Using the line plot below, place a dot at $11\frac{1}{2}$.

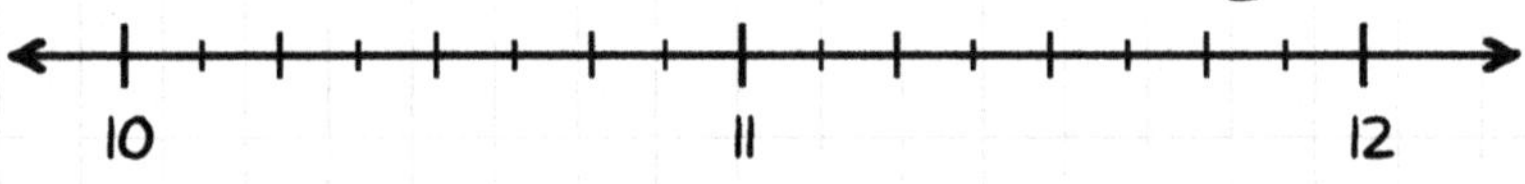

SHOW YOUR WORK

6. Which number could not be placed on the line plot below?

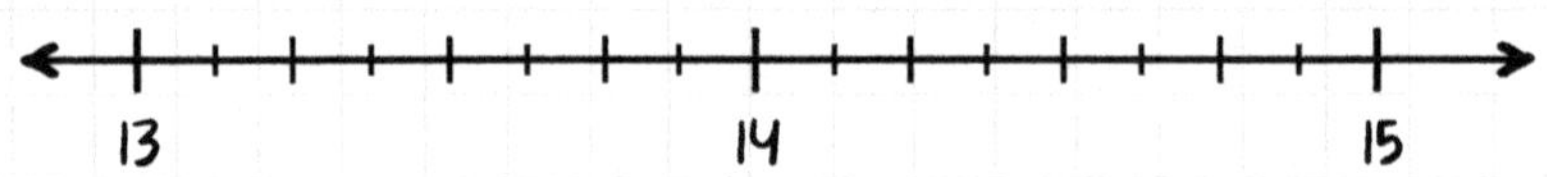

A. $13\frac{1}{2}$

B. $14\frac{7}{8}$

C. $15\frac{1}{2}$

D. $13\frac{3}{4}$

SHOW YOUR WORK

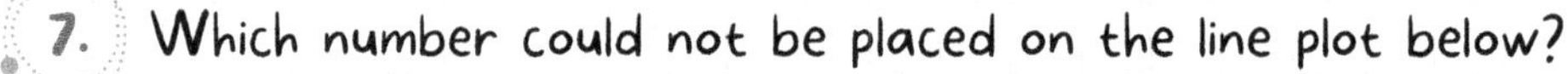

7. Which number could not be placed on the line plot below?

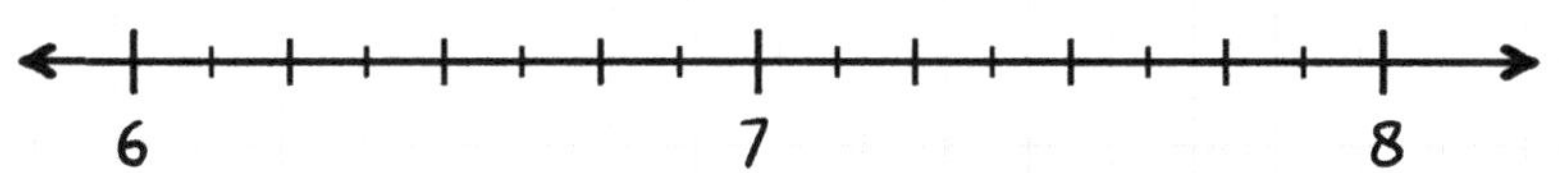

A. $6\frac{1}{8}$

B. $7\frac{1}{2}$

C. $7\frac{7}{8}$

D. $5\frac{3}{4}$

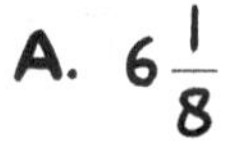

SHOW YOUR WORK

8. Which number could not be placed on the line plot below?

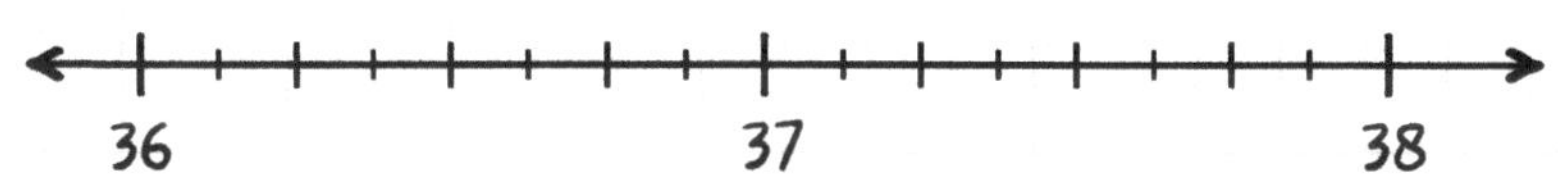

A. $35\frac{2}{4}$

B. $37\frac{7}{8}$

C. $36\frac{3}{4}$

D. $37\frac{1}{8}$

SHOW YOUR WORK

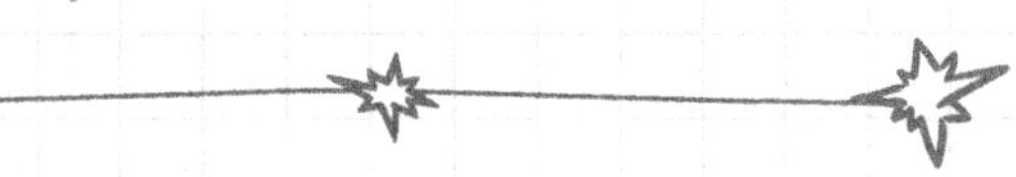

9. Which number could not be placed on the line plot below?

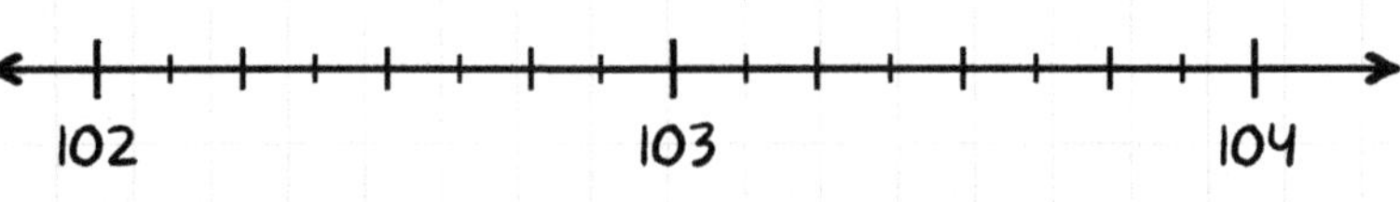

A. $103\frac{1}{4}$

B. $104\frac{3}{4}$

C. $102\frac{1}{8}$

D. $103\frac{2}{8}$

SHOW YOUR WORK

10. Which number could not be placed on the line plot below?

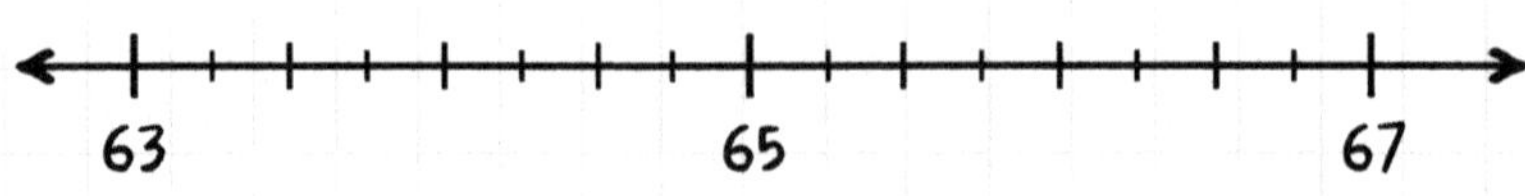

A. 64

B. 63

C. 66

D. 68

4.3.A | Understand concepts of angle and measure angles.

We can study measurement in terms of geometry too. Do you remember how we can divide shapes into lines and angles? **Angles can be measured by the center of the common endpoint of two rays.**

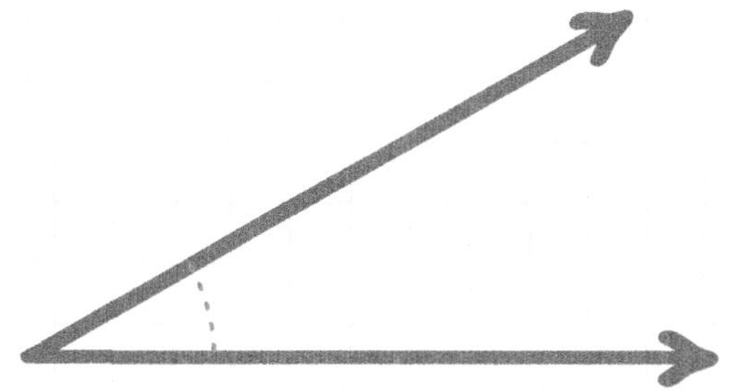

As angles get larger, they get closer to being a corner. The measurement of a corner angle or right angle is 90°. As angles grow larger than a right angle and approach becoming a straight line, they reach 180°. Angles that complete a circle measure a full 360°.

1. Which angle is largest?

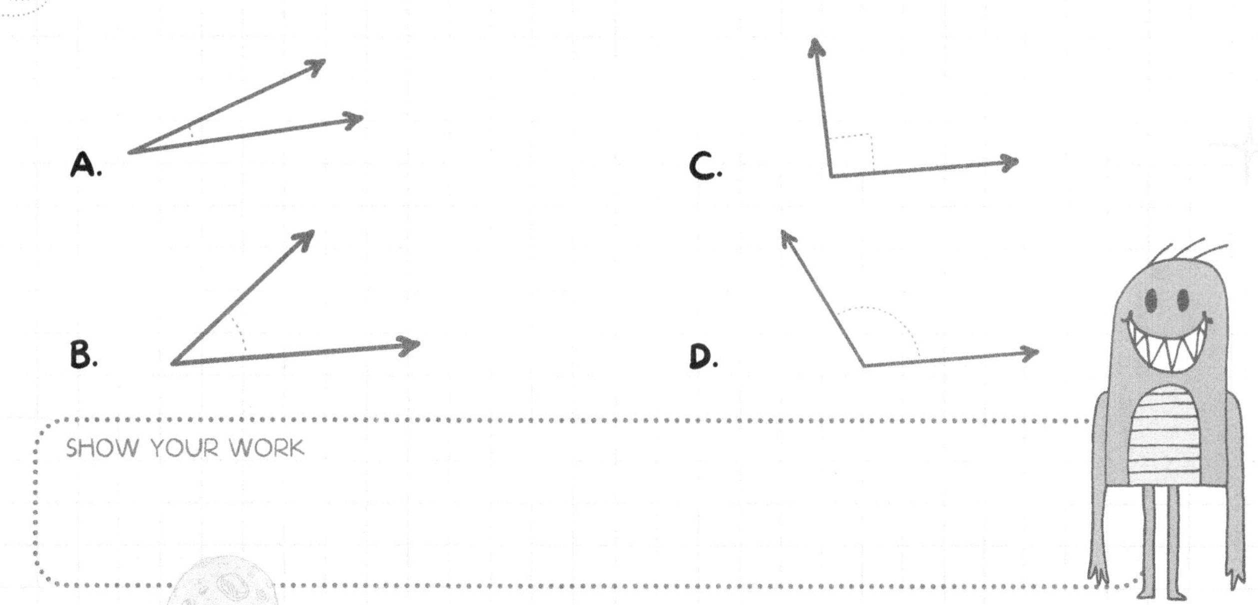

4.3.A Understand concepts of angle and measure angles.

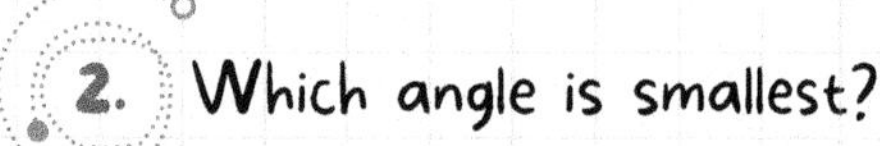

2. Which angle is smallest?

A. 

C. 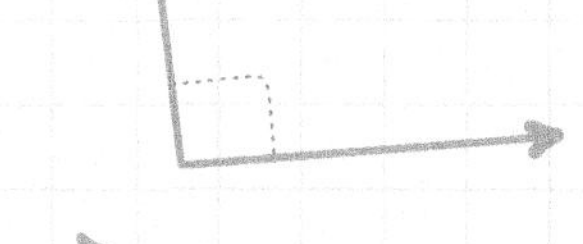

B.

D.

SHOW YOUR WORK

3. Which angle is a right angle?

A. 

C. 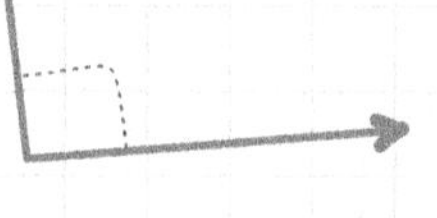

B. 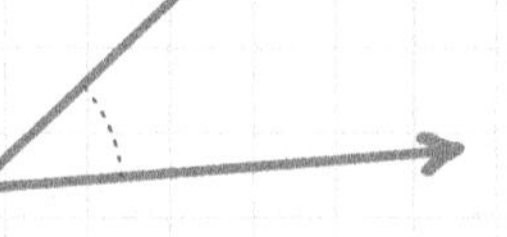

D.

SHOW YOUR WORK

4. Which angle is closest to 180°?

A. 

C. 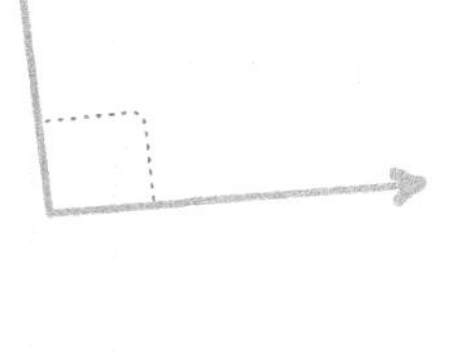

B.

D.

SHOW YOUR WORK

5. Draw an angle that is smaller than 90°.

SHOW YOUR WORK

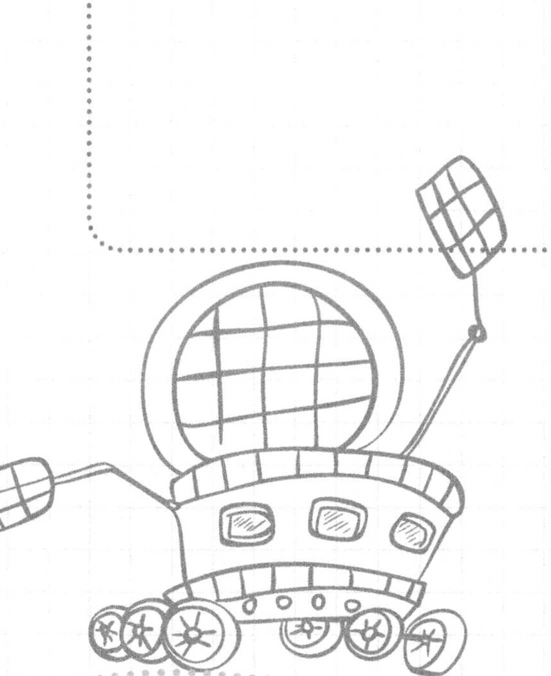

Angles are measured in terms of degrees. Each degree represents one part of an entire circle that you can imagine the shape to be a part of.

For example:

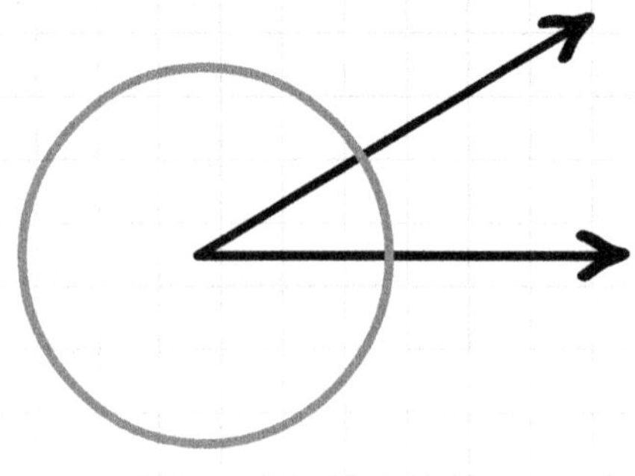

The angle is measured by its degrees of the circle.

Angles are named by their three points in order. One point on each end of the ray and the middle point in the middle.

For example, the angle below would be named Angle ABC.

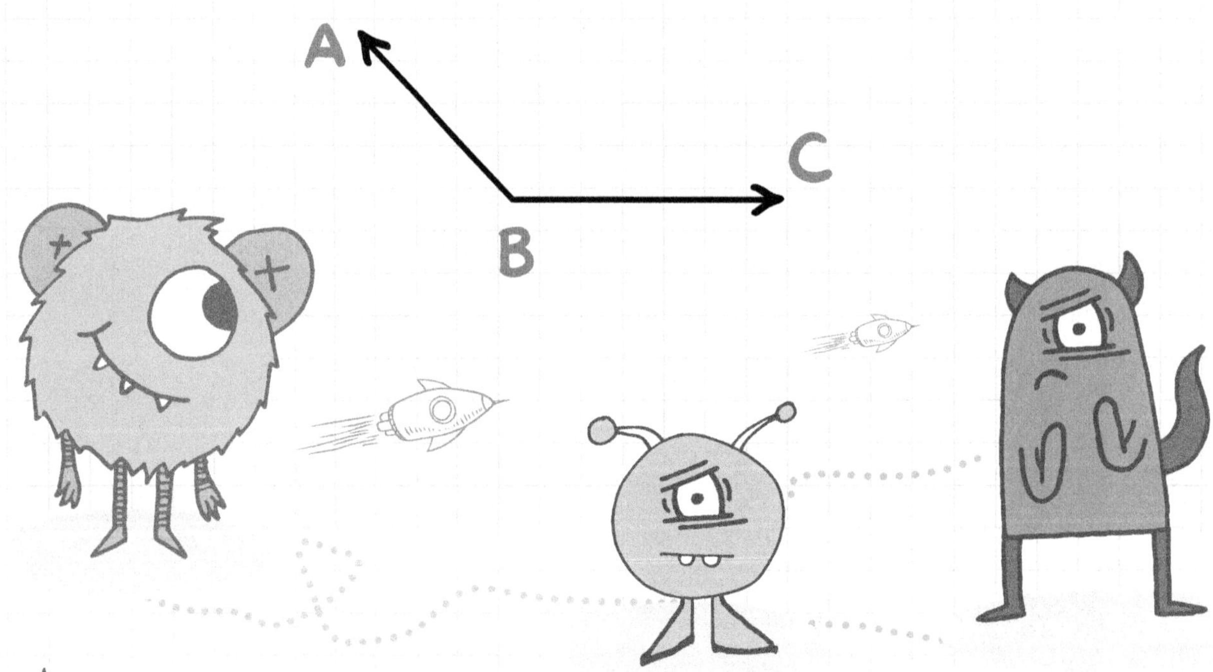

4.3.B | Understand concepts of angle and measure angles.

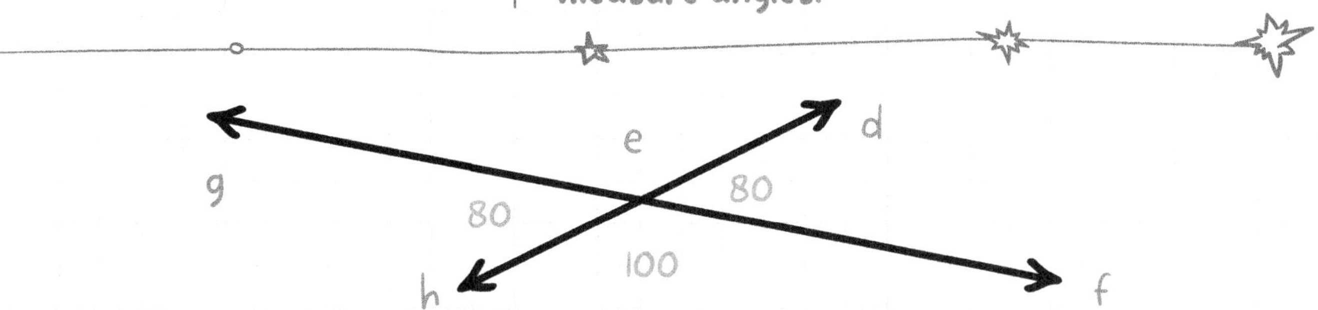

Use the picture above to answer questions 1-3.

1. Which angle is angle DEF?

A. 80° **B.** 180° **C.** 100° **D.** 90°

SHOW YOUR WORK

2. Which angle is angle DEH?

A. 80° **B.** 180° **C.** 100° **D.** 90°

SHOW YOUR WORK

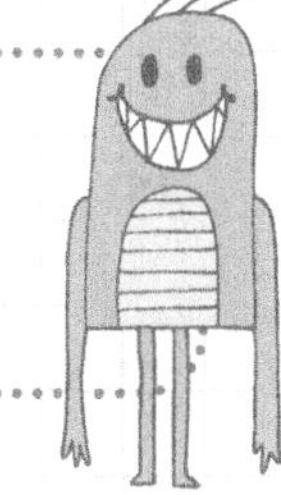

3. Which angle is angle HEF?

A. 80° **B.** 180° **C.** 100° **D.** 90°

SHOW YOUR WORK

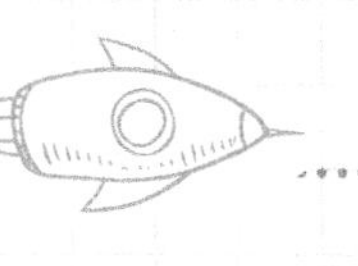

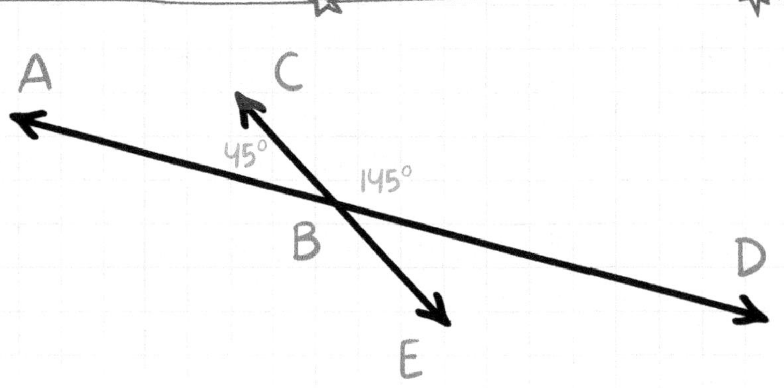

Use the picture below to answer questions 4 and 5.

4. Which angle has a measure of 45°?

A. CBE

B. ABC

C. BCA

D. ABD

SHOW YOUR WORK

5. Which angle has a measure of 145°?

A. CBD

B. ABC

C. ECB

D. BDE

SHOW YOUR WORK

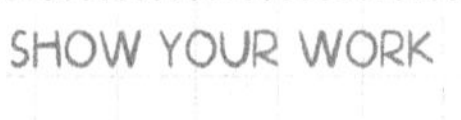

4.3.C | Understand concepts of angle and measure angles.

How do you find an angle measurement you do not know? One way to to use a protractor. A protractor is a math tool specifically designed to measure angles. Protractors look like this:

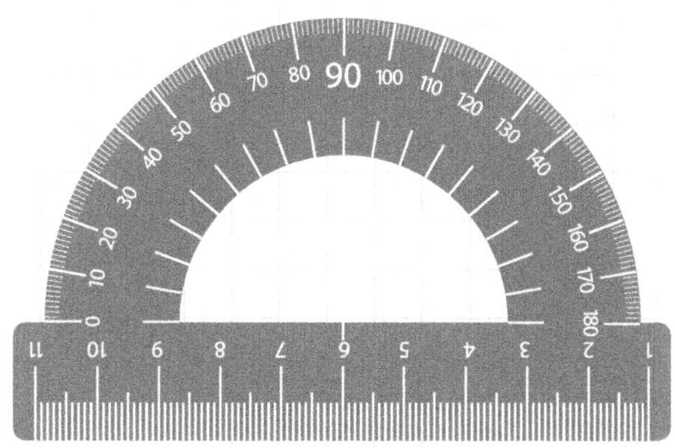

Using protractors are easy! First you line the middle of the protractor up with the middle of the angle. You line one ray up with 0° and read the measurement of the other ray.

You can also use protractors to draw angles of a certain measurement. You put a point on your paper. You line the point up with the middle of the protractor. Then you draw a dot at the 0° measurement and then a dot at the angle you are looking for. You connect the dots with your original points.

1. Use a protractor to measure the following angle and choose the closest measurement.

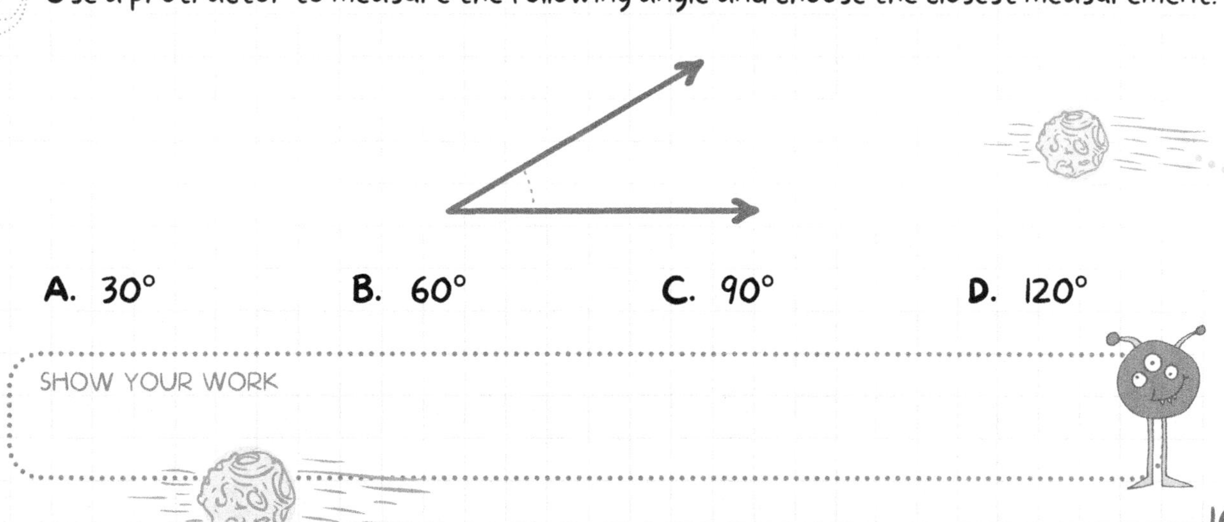

A. 30° **B.** 60° **C.** 90° **D.** 120°

SHOW YOUR WORK

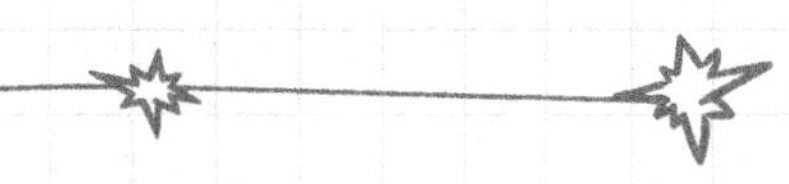

2. Use a protractor to measure the following angle and choose the closest measurement.

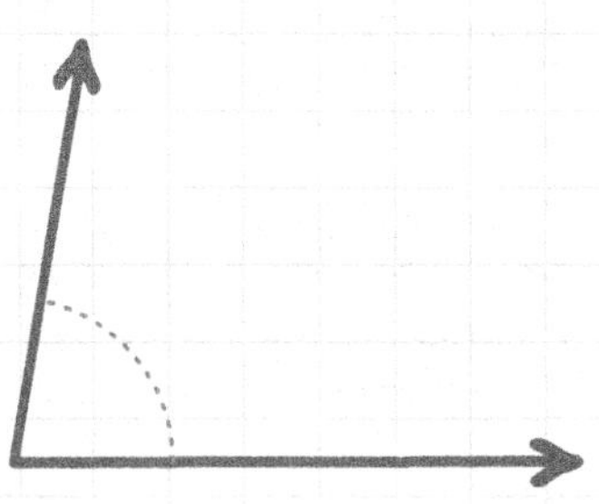

A. 30°　　B. 45°　　C. 80°　　D. 110°

SHOW YOUR WORK

3. Use a protractor to measure the following angle and choose the closest measurement.

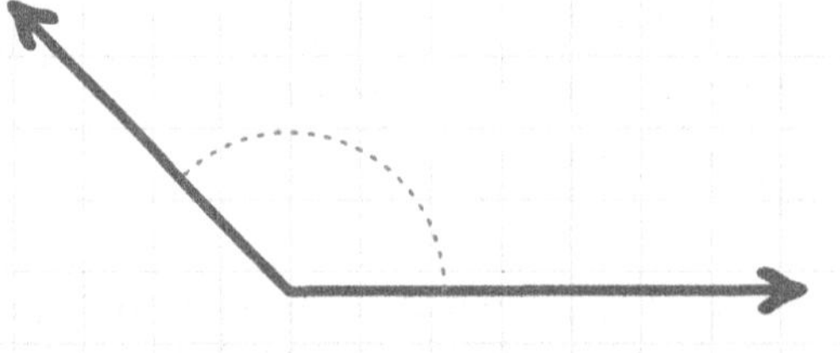

A. 45°　　B. 85°　　C. 100°　　D. 135°

SHOW YOUR WORK

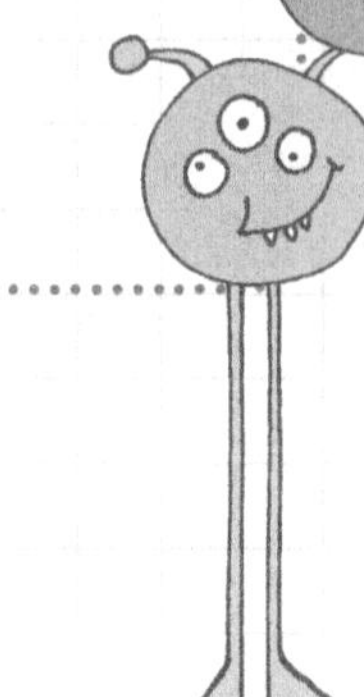

4. Use a protractor to draw an angle that measures around 45°.

SHOW YOUR WORK

5. Use a protractor to draw an angle that measures around 100°.

SHOW YOUR WORK

NOTES

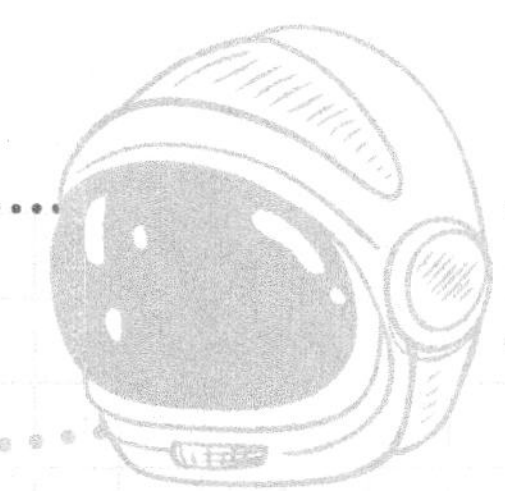

4.3.D | Understand concepts of angle and measure angles.

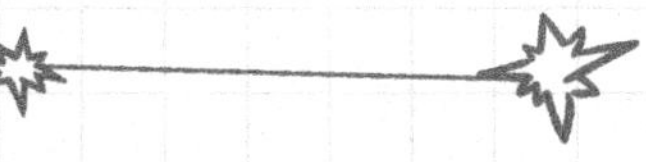

We can combine angle measurements and add angles. When an angle is part of a circle, we can add additional angles to make the original angle larger. We can also work backward and subtract angle measurements to get a smaller angle.

1. If angle mop = 80°, which equation tells us the measure of angle nop?

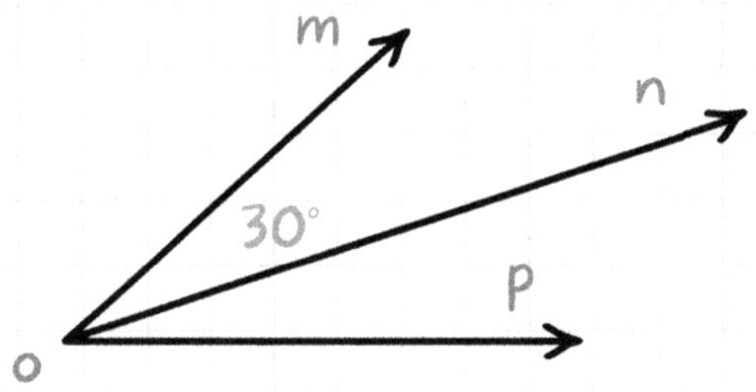

A. $80 - 30 = x$

B. $80 + 30 = x$

C. $30 + 30 = x$

D. $x + 80 = 30$

SHOW YOUR WORK

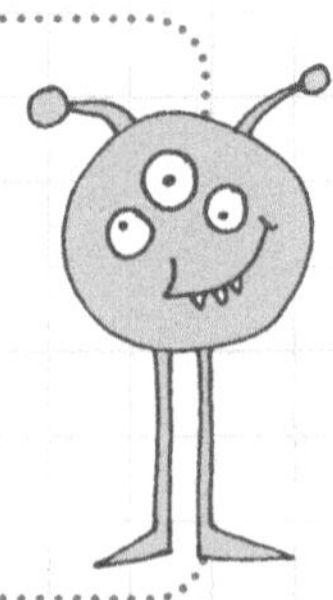

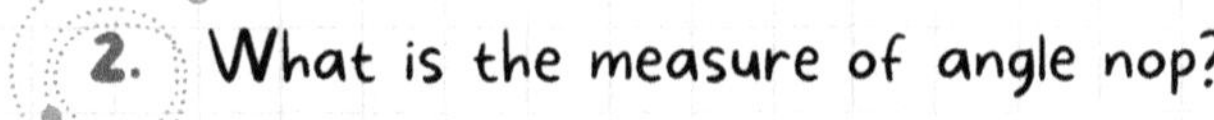

2. What is the measure of angle nop?

A. 45°

B. 48°

C. 50°

D. 110°

SHOW YOUR WORK

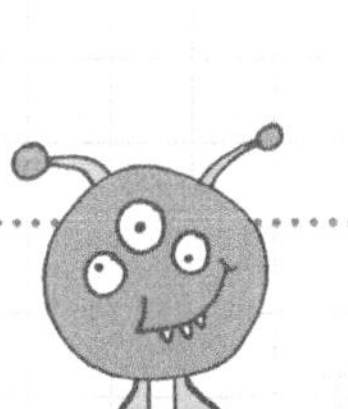

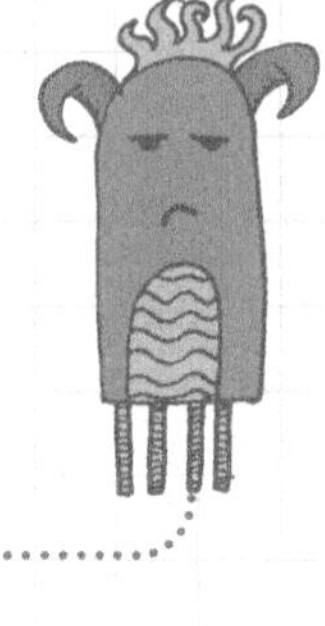

3. If angle d is 110° and it is divided into two angles, a right angle and angle b. What is the measure of angle b?

A. 10°

B. 20°

C. 30°

D. 40°

SHOW YOUR WORK

4. If the clock is set at 6:30 and the minute hand moves 90°, what is the new time on the clock?

A. 6 : 00

B. 7 : 00

C. 6 : 45

D. 7 : 30

SHOW YOUR WORK

5. If the clock is set at 1:15 and the minute hand moves 180°, what is the new time on the clock?

A. 1 : 00

B. 1 : 30

C. 1 : 45

D. 2 : 15

SHOW YOUR WORK

4.4. Chapter Test

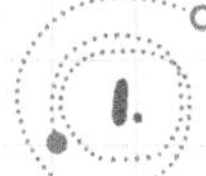

1. What measurement completes the table below?

kilograms	grams
7.5	x

A. 750

B. 7000

C. 500

D. 7500

2. Which measurement is equal to 40 minutes?

A. 240 hours

B. 24 hours

C. 2400 seconds

D. 24 seconds

3. Which measurement is equal to 632 cm?

A. 0.632 m

B. 63.2 m

C. 6.32 m

D. 632 m

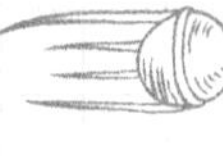

4. Eleanor uses the following recipe to make lemonade: 4 pints of lemon juice, 4 cups of sugar syrup, 6 quarts of water. If 2 cups = 1 pint and 2 pints = 1 quart, how much liquid will Eleanor have once she makes the lemonade?

A. 30 cups

B. 36 cups

C. 24 cups

D. 28 cups

SHOW YOUR WORK

5. Lidiya is 4 feet 8 inches tall. Her sister is 6 inches taller. How tall is Lidiya's sister?

A. 5 feet

B. 5 feet 2 inches

C. 5 feet 12 inches

D. 5 feet 6 inches

SHOW YOUR WORK

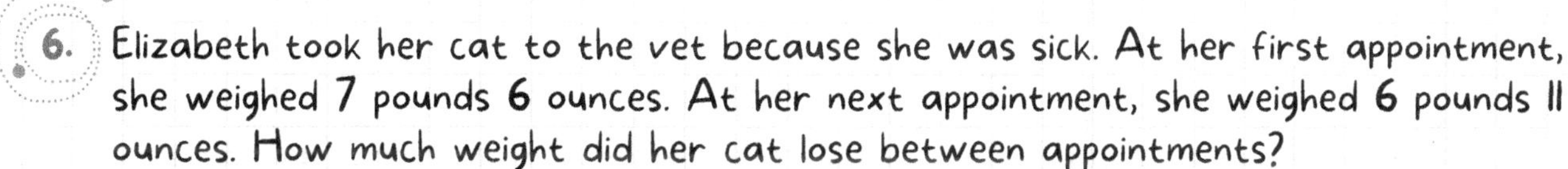

6. Elizabeth took her cat to the vet because she was sick. At her first appointment, she weighed 7 pounds 6 ounces. At her next appointment, she weighed 6 pounds 11 ounces. How much weight did her cat lose between appointments?

A. 1 pound

B. 15 ounces

C. 11 ounces

D. 14 ounces

SHOW YOUR WORK

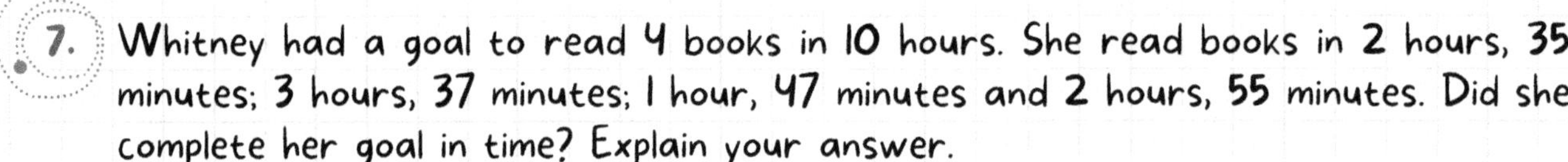

7. Whitney had a goal to read 4 books in 10 hours. She read books in 2 hours, 35 minutes; 3 hours, 37 minutes; 1 hour, 47 minutes and 2 hours, 55 minutes. Did she complete her goal in time? Explain your answer.

SHOW YOUR WORK

8. Ryan likes to go to the climbing wall at his gym. When he started climbing, he made it 20 feet, 8 inches. After three weeks, he was able to climb up to 28 feet, 6 inches. How much did Ryan improve his climbing skill?

A. 7 feet 8 inches

B. 7 feet

C. 7 feet 4 inches

D. 7 feet 10 inches

SHOW YOUR WORK

9. The perimeter of a garden is measured as 68 ft. The length of the garden is 18 feet. How long is the width?

A. 16 feet

B. 17 feet

C. 18 feet

D. 19 feet

SHOW YOUR WORK

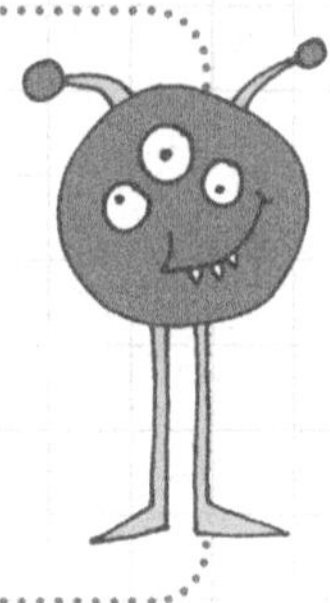

4.4. Chapter Test

10. If a rectangle has a width of **5** and a perimeter of **26**, what is the area?

A. 35

B. 40

C. 45

D. 50

SHOW YOUR WORK

11. Using the line plot below, where would you graph $3\frac{5}{8}$?

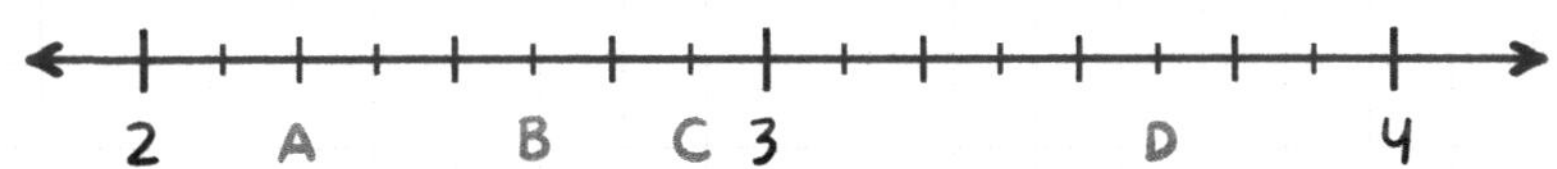

A. Point A

B. Point B

C. Point C

D. Point D

SHOW YOUR WORK

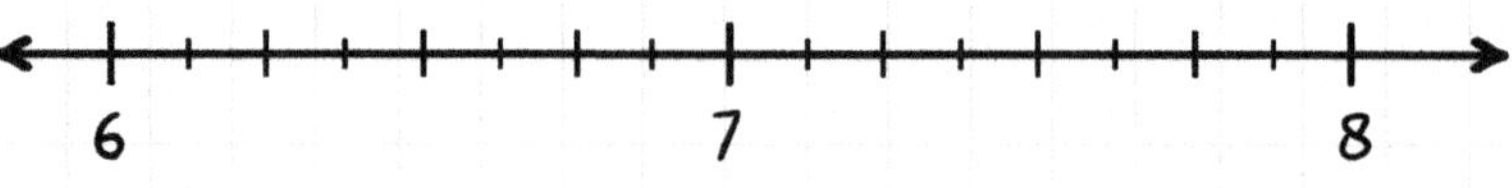

12. Which number could not be placed on the line plot below?

6 7 8

A. $6\frac{1}{8}$

B. $7\frac{1}{2}$

C. $5\frac{7}{8}$

D. $7\frac{3}{4}$

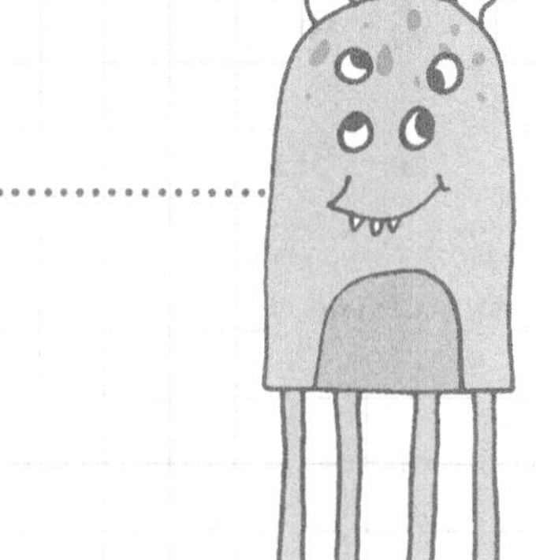

SHOW YOUR WORK

13. Which mixed number is the smallest?

A. $35\frac{2}{4}$

B. $37\frac{7}{8}$

C. $36\frac{3}{4}$

D. $37\frac{1}{8}$

SHOW YOUR WORK

14. Draw an angle that is larger than 90° and smaller than 180°.

SHOW YOUR WORK

4.4. Chapter Test

15. Which angle is angle TUV?

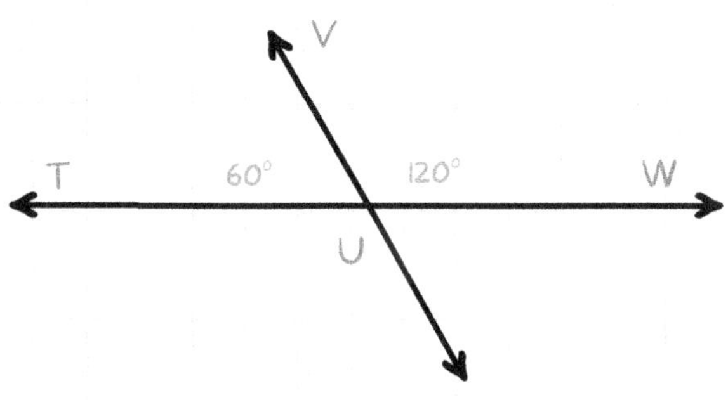

A. 45° **B.** 50° **C.** 60° **D.** 180°

SHOW YOUR WORK

16. Which angle has a measure of 120°?

A. uwv

B. vuw

C. tuv

D. vut

SHOW YOUR WORK

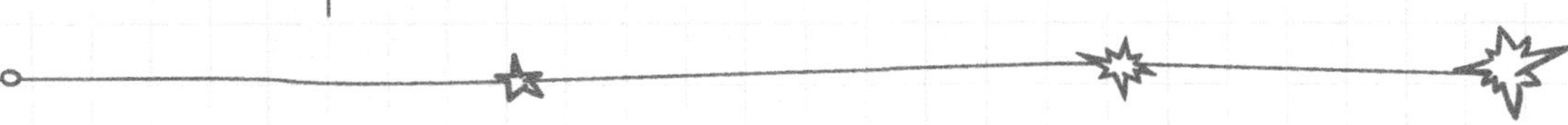

17. Use a protractor to measure the following angle and choose the closest measurement.

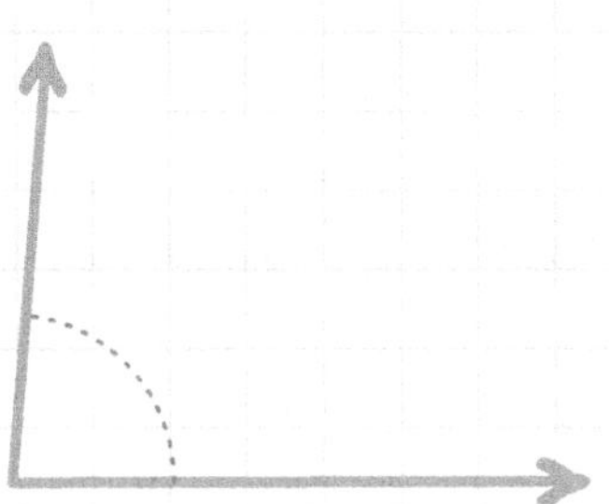

A. 70°

B. 75°

C. 80°

D. 85°

18. Use a protractor to draw an angle that measures around 110°.

SHOW YOUR WORK

19. Angle x is 100° and it is divided into two angles, a and b (both the same size). What is the measure of angle b?

A. $b = 100 - 2$

B. $b = 100 \div 2$

C. $b = 100 \times 2$

D. $b = 100 + 100$

SHOW YOUR WORK

20. If the clock is set at 4 : 00 and the minute hand moves 180°, what is the new time on the clock?

A. 4 : 30

B. 4 : 45

C. 5 : 00

D. 4 : 15

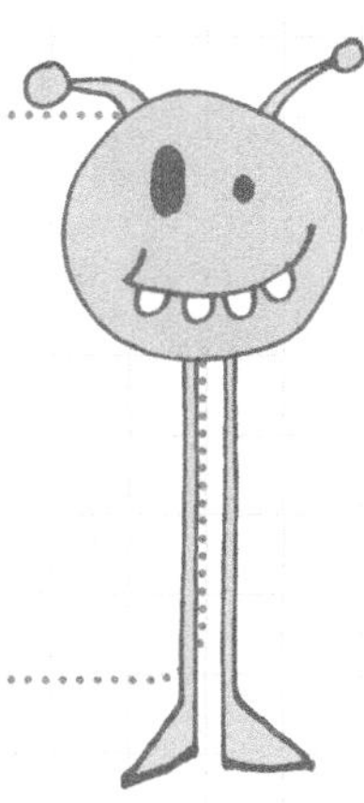

Chapter 5 : Geometry

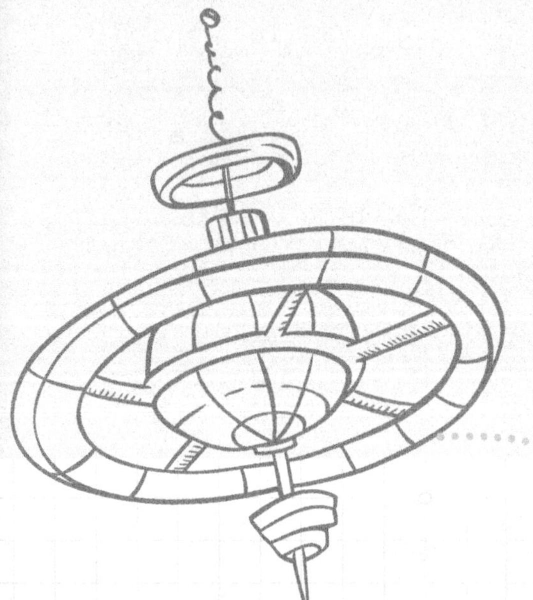

5.1.A Draw and identify lines and angles, and classify shapes by properties of their lines and angles

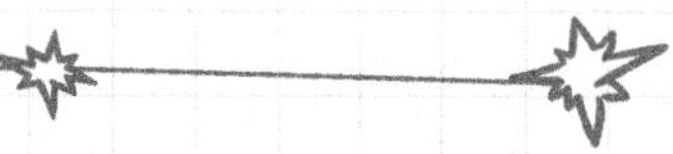

We can use our knowledge of geometry terms to analyze lines and properties of lines.

Review the vocabulary terms below and then apply those to the practice problems.

Point: Represented by a dot, it has no length or width

Line: A line of points that extends forever in both directions (represented by arrows on both sides)

Line segment: A part of a line that has a beginning and an end point

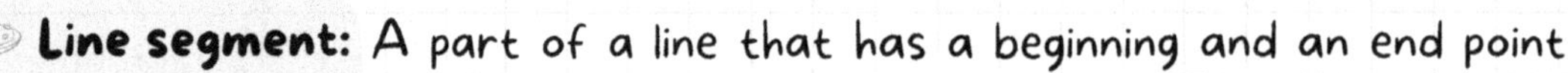

Ray: A part of a line that has a beginning but no end point (represented with an arrow on one side)

Angle: Two rays that come together to form a measurement less than 360°

Right angle: An angle that measures 90°

Acute angle: An angle that is smaller than 90°

Obtuse angle: An angle that is larger than 90°

Perpendicular lines: Two lines that intersect and form a 90° angle

Parallel lines: Two lines that never intersect

5.1.A

Draw and identify lines and angles, and classify shapes by properties of their lines and angles

1. Identify the line.

A.

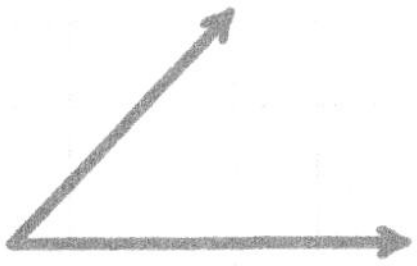

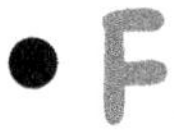

B.

C. • F

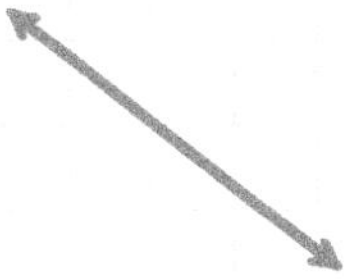

D.

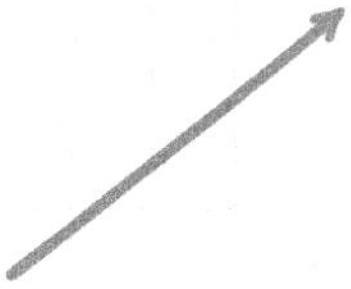

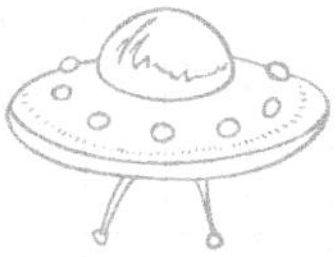

SHOW YOUR WORK

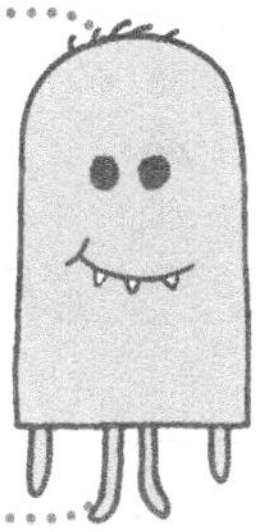

2. Identify the ray.

A.

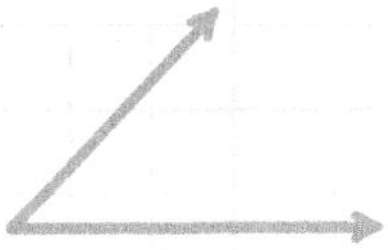

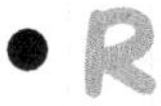

B.

C. • R

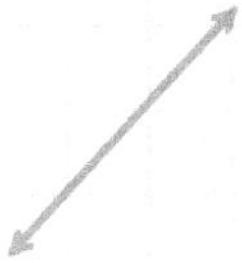

D.

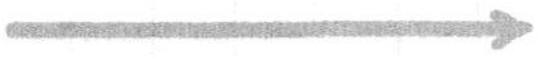

SHOW YOUR WORK

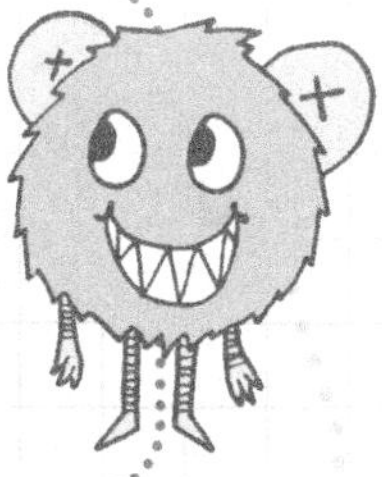

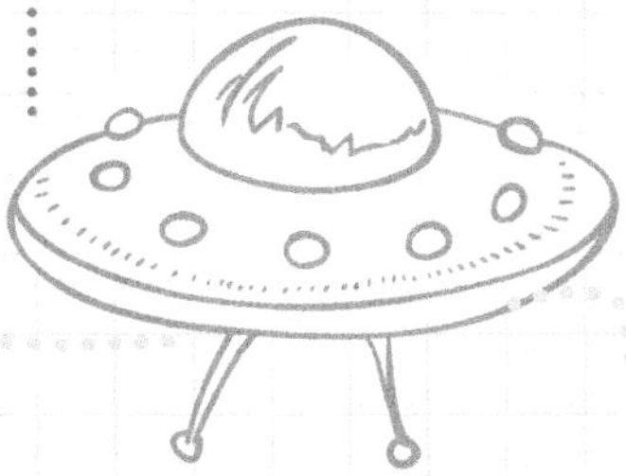

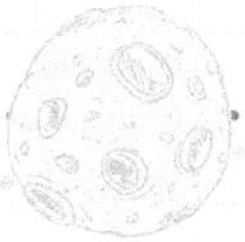

3. Draw a pair of perpendicular lines.

SHOW YOUR WORK

4. Which angle is acute?

A.

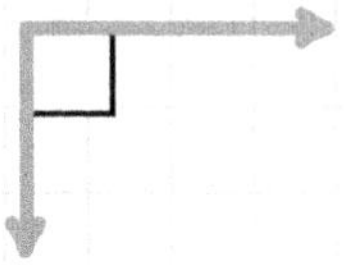

B.

C.

D.

SHOW YOUR WORK

5. Which angle is right?

A.

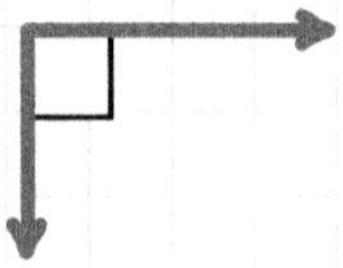

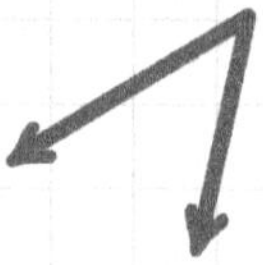

B.

C.

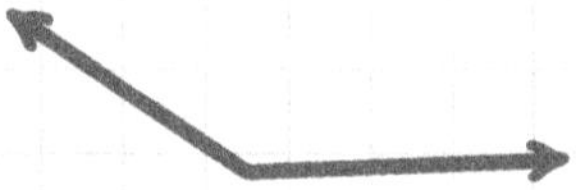

D.

SHOW YOUR WORK

5.1.A

Draw and identify lines and angles, and classify shapes by properties of their lines and angles

6. Draw an obtuse angle.

SHOW YOUR WORK

7. In the shape below, what is the point?

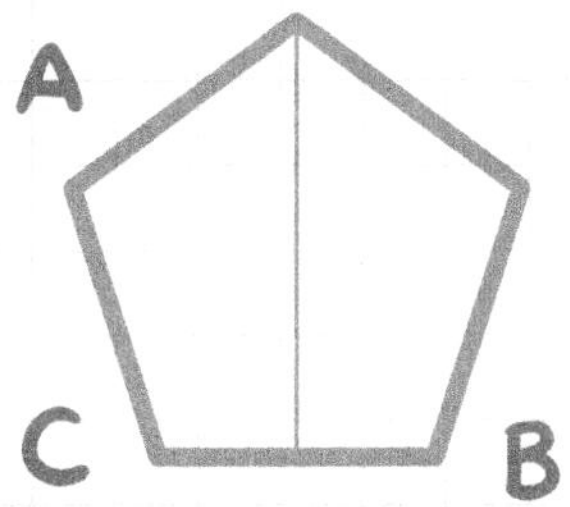

SHOW YOUR WORK

A. Side A

B. Point C

C. Line segment CB

D. Angle Bd

8. In the shape below, circle the parallel lines.

SHOW YOUR WORK

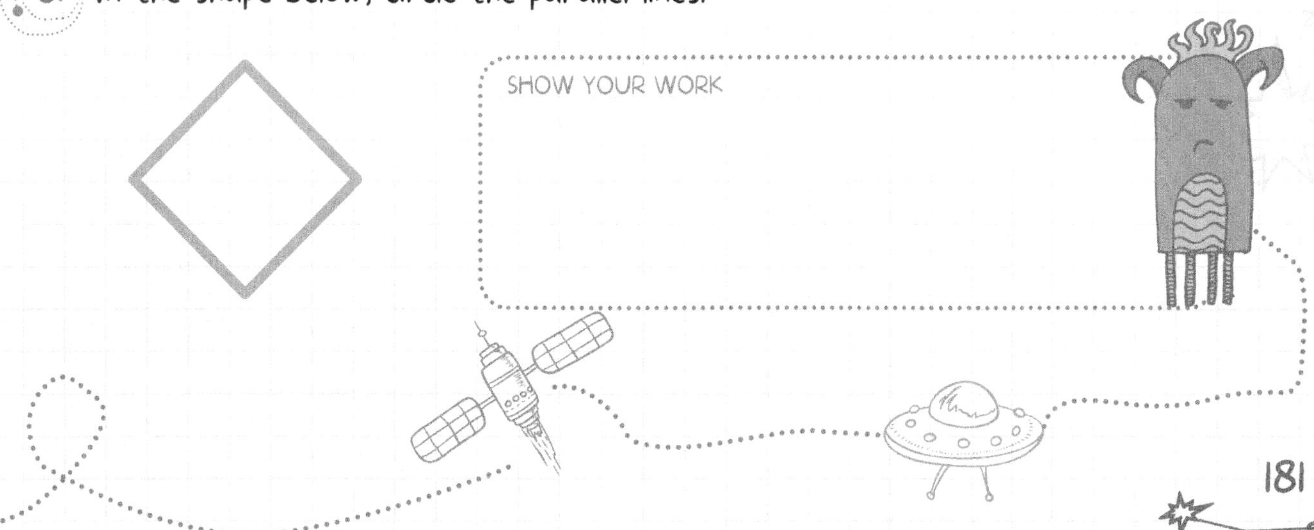

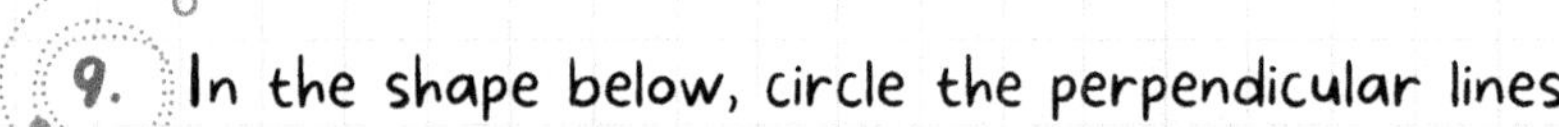

9. In the shape below, circle the perpendicular lines.

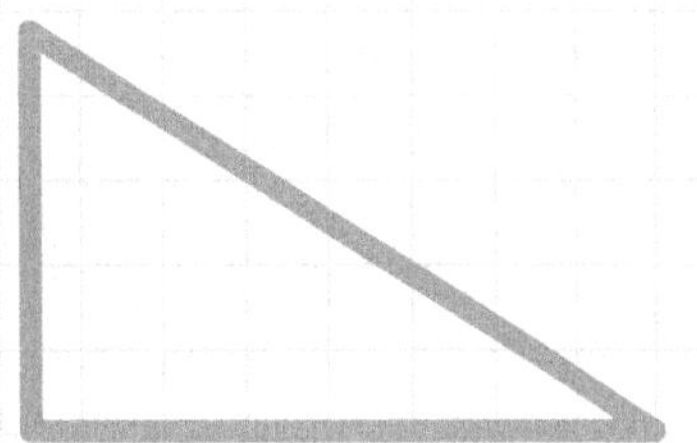

SHOW YOUR WORK

10. Draw a shape that includes 1 set of perpendicular and 1 set of parallel lines.

SHOW YOUR WORK

NOTES

5.1.B

Draw and identify lines and angles, and classify shapes by properties of their lines and angles

We can apply our knowledge of parallel and perpendicular lines to shapes. Some shapes contain lines that are parallel, some shapes contain lines that are perpendicular, some shapes contain both types of lines and some shapes contain neither type.

We can also use our knowledge of angles to analyze shapes. **Shapes with perpendicular lines contain right angles.**

Triangles have special properties. The three angles in a triangle always add up to 180°. This leads to a few special types of triangles:

- **Right triangles** contain one angle that is 90°.
- **Equilateral triangles** contain three congruent or equal angles. Because all triangles add up to 180°, the measurements of angles in an equilateral triangle always add up to 60°.
- **Acute triangles** contain three acute angles, or angles less than 90°.
- **Obtuse triangles** contain one angle that is larger than 90°.
- **Scalene triangles** have sides that are all different lengths.
- **Isosceles triangles** have two equal sides and two equal angles.

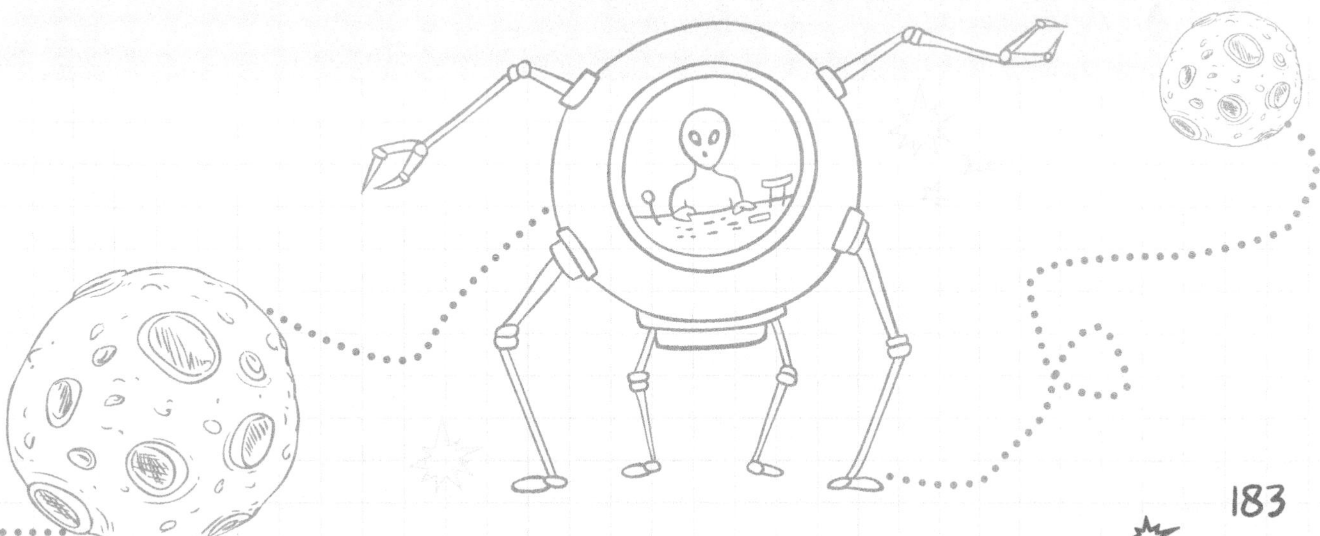

1. Which shape contains a set of perpendicular lines?

A. An equilateral triangle

B. A right triangle

C. An acute triange

D. An obtuse triangle

SHOW YOUR WORK

2. Which shape contains 2 sets of parallel lines?

A. A trapezoid

B. A triangle

C. A pentagon

D. A hexagon

SHOW YOUR WORK

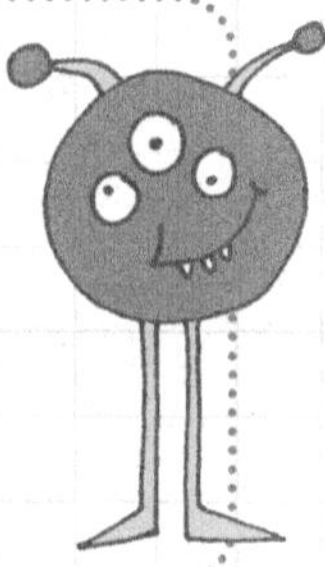

3. Which shape contains both perpendicular and parallel lines?

A. A trapezoid

B. A triangle

C. A rectangle

D. A hexagon

SHOW YOUR WORK

4. Which shape does not contain perpendicular or parallel lines?

A. A trapezoid

B. A scalene triangle

C. A diamond

D. A right triangle

SHOW YOUR WORK

5. Which triangle's angle measurements represent a right triangle?

A. 90°, 40°, 50°

B. 100°, 20°, 60°

C. 60°, 60°, 60°

D. 50°, 50°, 80°

SHOW YOUR WORK

6. Which triangle's angle measurements represent an obtuse triangle?

A. 90°, 40°, 50°

B. 100°, 20°, 60°

C. 60°, 60°, 60°

D. 50°, 50°, 80°

SHOW YOUR WORK

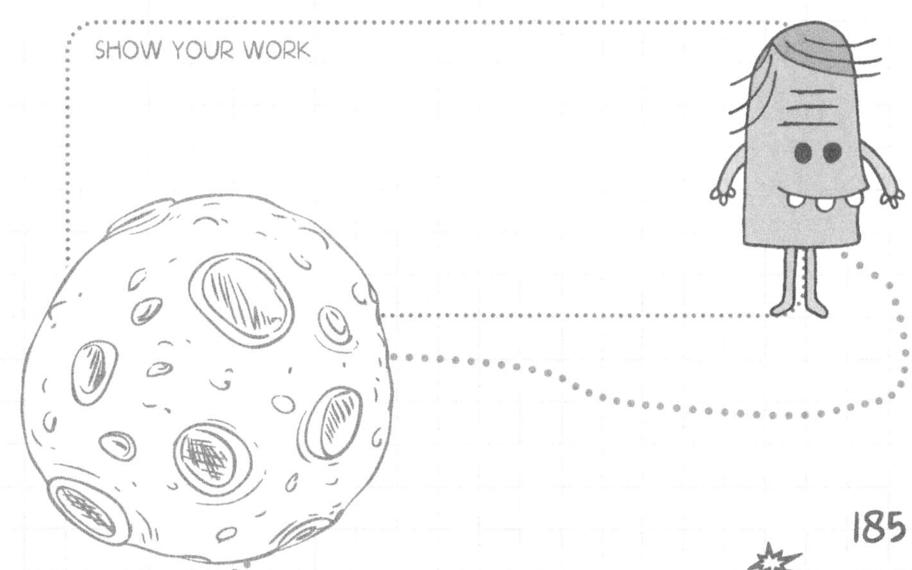

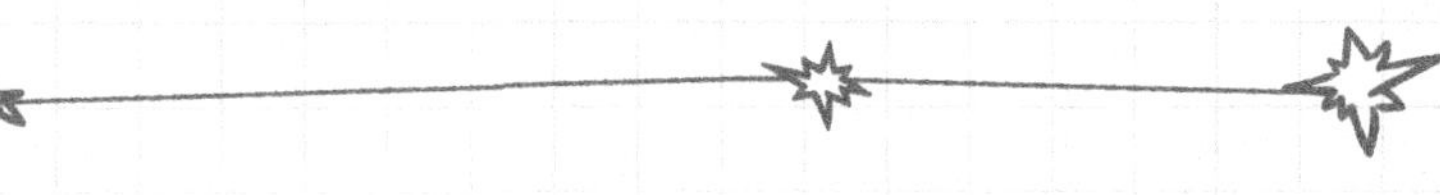

7. How do you know this triangle is scalene?

A. It has a right angle.

B. It has an obtuse angle.

C. It has two equal sides.

D. It has three unequal sides.

SHOW YOUR WORK

8. What are the angle measurements of an equilateral triangle?

A. 50°, 50°, 50°

B. 75°, 75°, 75°

C. 60°, 60°, 60°

D. 100°, 100°, 100°

SHOW YOUR WORK

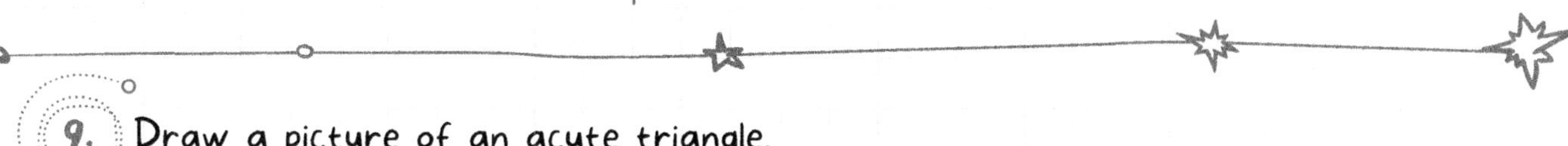

9. Draw a picture of an acute triangle.

SHOW YOUR WORK

10. Draw a picture of an isosceles triangle.

SHOW YOUR WORK

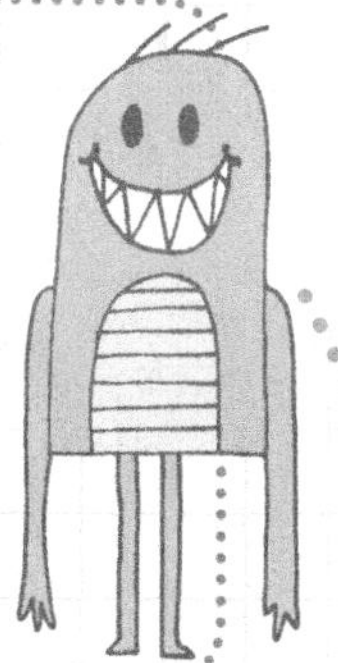

NOTES

We can also analyze two-dimensional shapes based on their lines of symmetry. **A line of symmetry is a line we can draw across a figure and the shape can be broken along the line into equal parts.**

This is a shape with a line of symmetry.

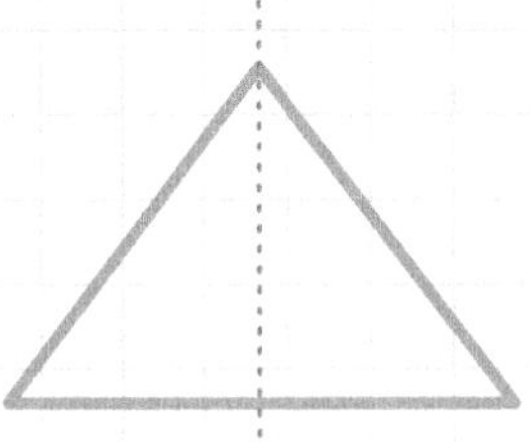

If we fold the triangle along the line, we would have two equal parts.

This is a shape that does not have a line of symmetry.

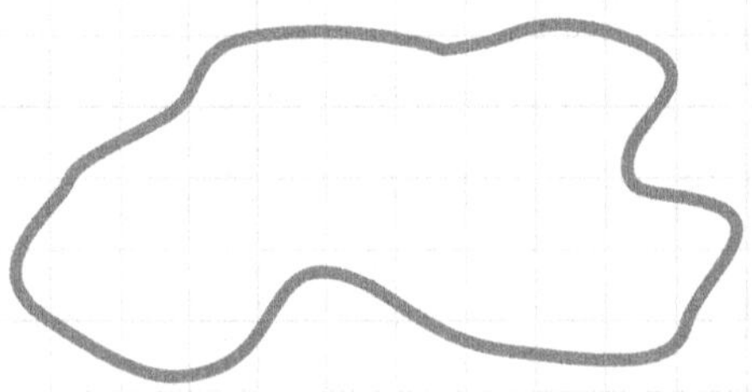

There is no line I can draw anywhere along that shape that will divide it evenly.

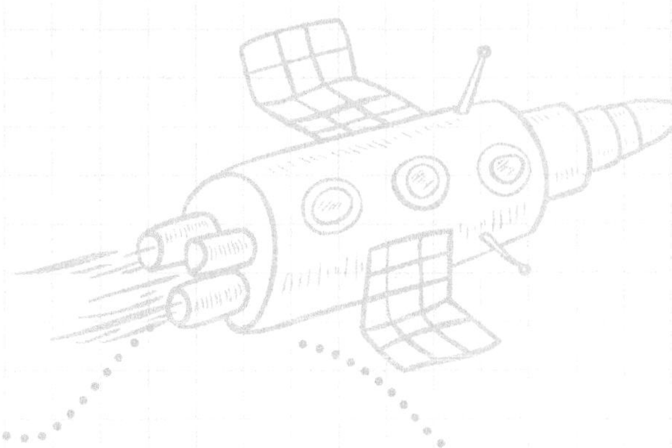

5.1.C Draw and identify lines and angles, and classify shapes by properties of their lines and angles

1. Draw a line of symmetry on the following shape.

SHOW YOUR WORK

2. Draw a line of symmetry on the following shape.

SHOW YOUR WORK

3. Which line of symmetry is correct?

A.

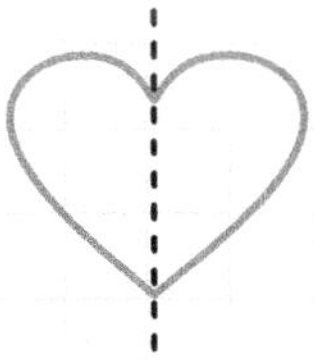

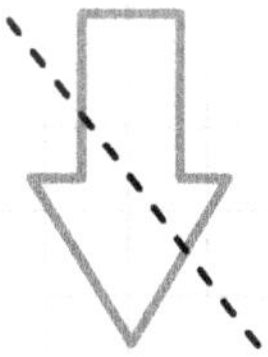

B.

C.

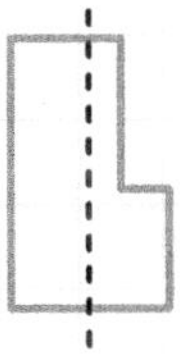

D.

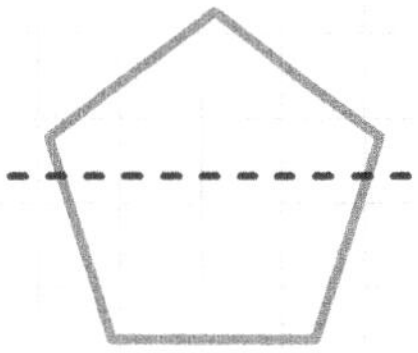

SHOW YOUR WORK

4. Which line of symmetry is correct?

A.

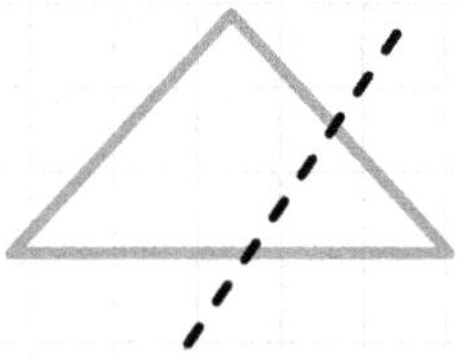

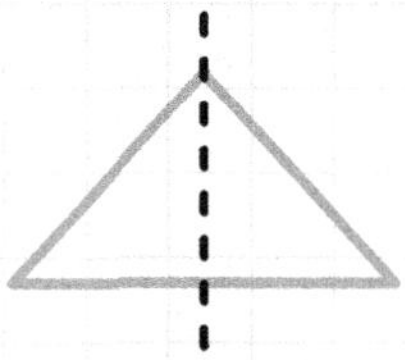

B.

C.

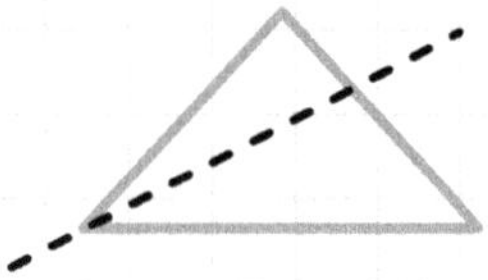

D.

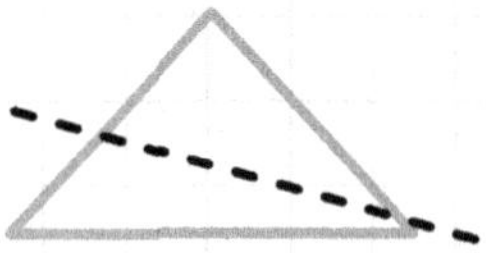

SHOW YOUR WORK

5. Which line of symmetry is correct?

A.

B.

C.

D.

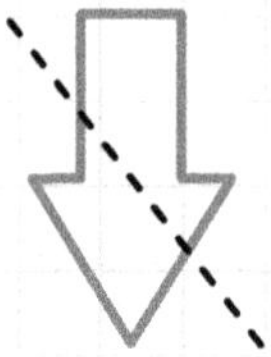

SHOW YOUR WORK

6. Draw a line of symmetry on the following shape.

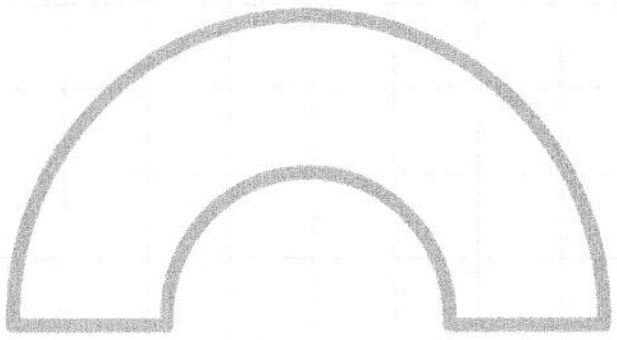

SHOW YOUR WORK

7. Which shape has a line of symmetry?

A.

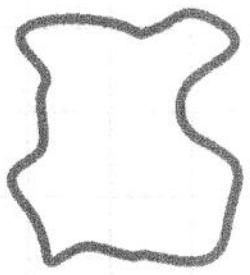

B.

C.

D.

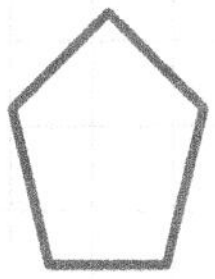

SHOW YOUR WORK

8. Which shape does not have a line of symmetry?

A.

B.

C.

D.

SHOW YOUR WORK

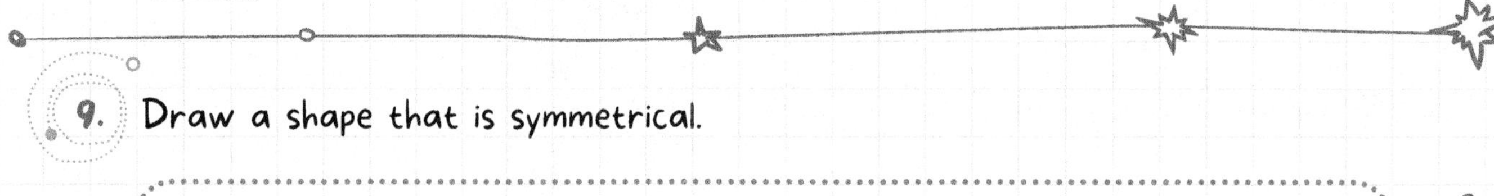

9. Draw a shape that is symmetrical.

SHOW YOUR WORK

10. Draw a shape that is not symmetrical.

SHOW YOUR WORK

5.2. Chapter Test

1. Identify the line segment.

A.

B.

C.

D.

SHOW YOUR WORK

2. Which angle is obtuse?

A.

C.

B.

D.

SHOW YOUR WORK

3. Draw an example of a point.

SHOW YOUR WORK

4. Which shape contains parallel lines?

A.

B.

C.

D.

SHOW YOUR WORK

5. Draw a shape that does not include perpendicular or parallel lines.

SHOW YOUR WORK

6. How would you identify this triangle?

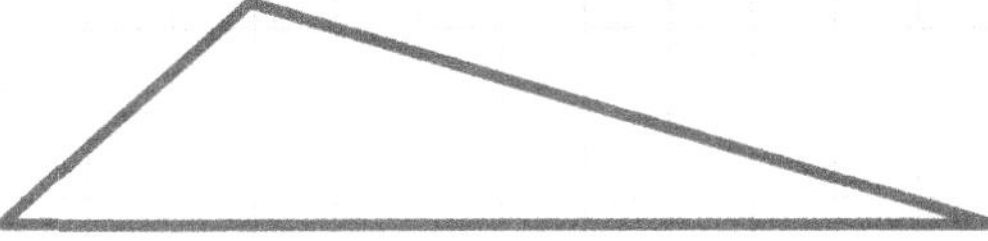

A. Isosceles

B. Right

C. Obtuse

D. Acute

SHOW YOUR WORK

7. How would you identify this triangle?

A. Isosceles

B. Right

C. Obtuse

D. Scalene

8. What is special about an equilateral triangle?

A. Equal angles

B. All angles 45°

C. Contains a right angle

D. Contains three different angles

SHOW YOUR WORK

9. Which triangle's angle measurements represent an acute triangle?

A. 60°, 30°, 90°

B. 70°, 50°, 60°

C. 90°, 45°, 45°

D. 120°, 35°, 25°

SHOW YOUR WORK

10. Draw a picture of a right triangle.

SHOW YOUR WORK

11. Draw a line of symmetry on the following shape.

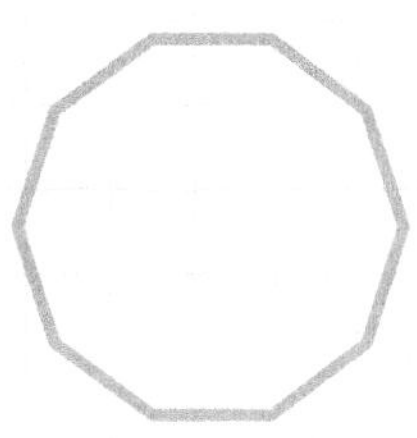

SHOW YOUR WORK

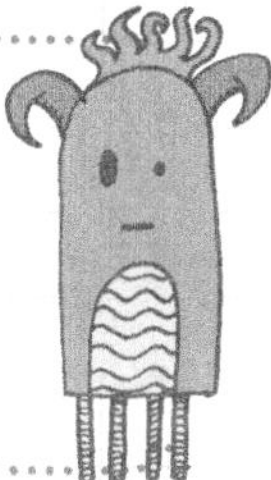

12. Which line of symmetry is correct?

A. 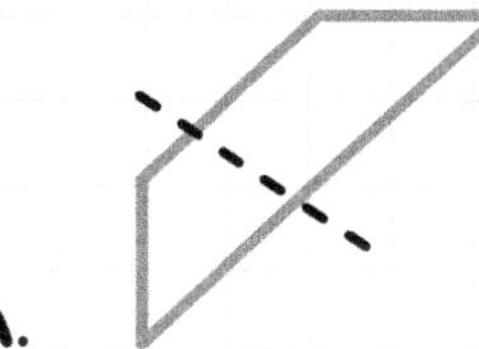

C. 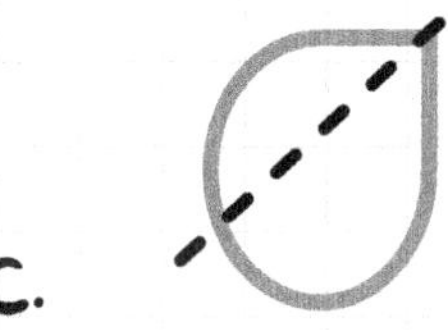

B. 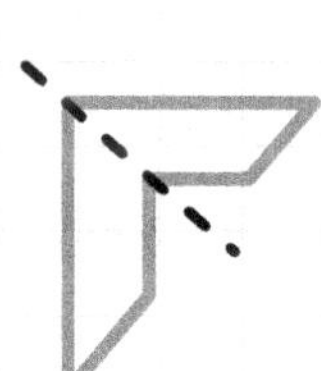

D. 

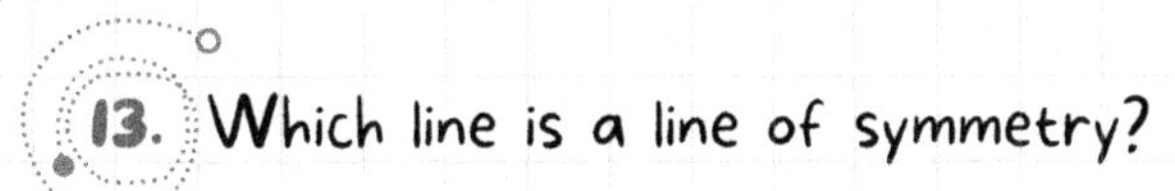

13. Which line is a line of symmetry?

A. 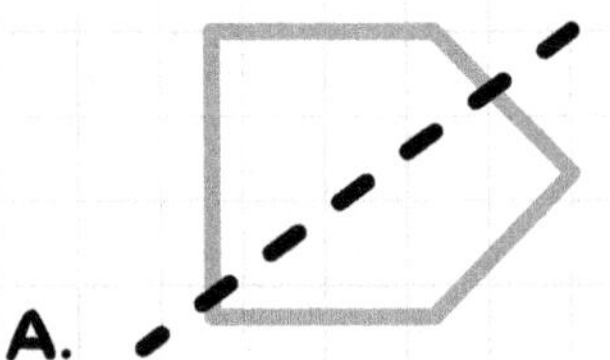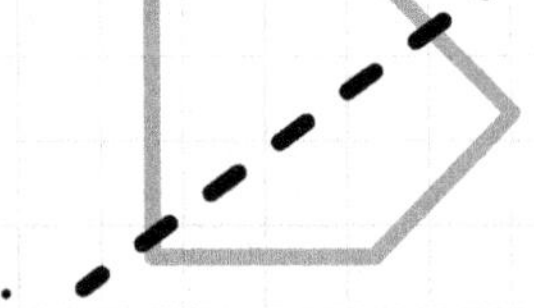

C. 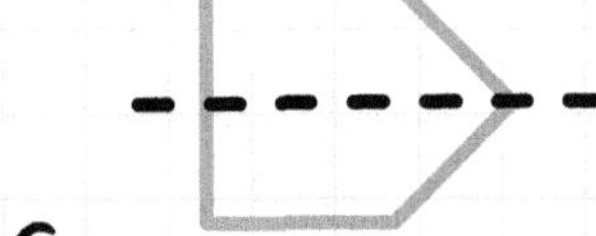

B. 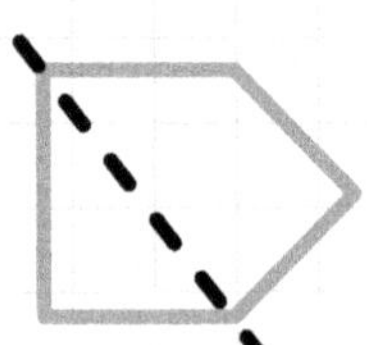

D. 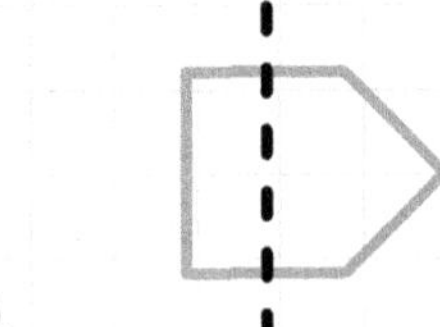

SHOW YOUR WORK

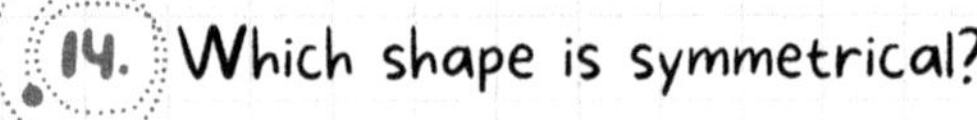

14. Which shape is symmetrical?

A. 

C. 

B.

D. 

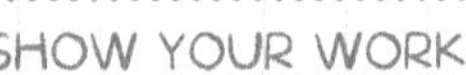

SHOW YOUR WORK

15. Which shape does not have a line of symmetry?

A. 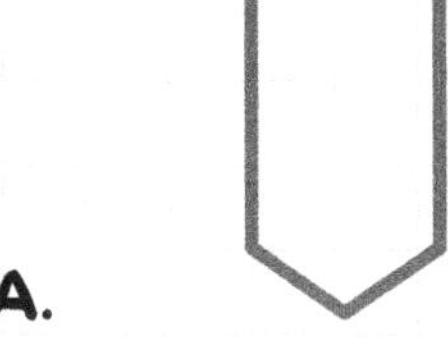

C. 

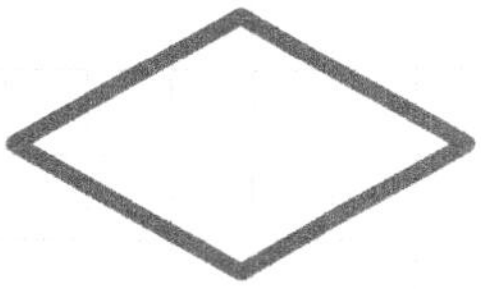

B. 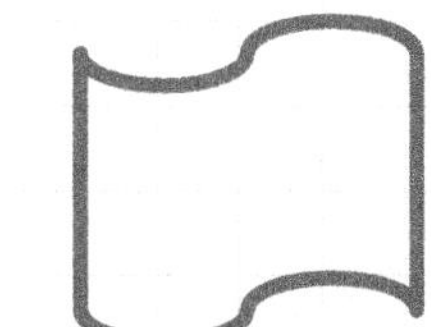

D.

SHOW YOUR WORK

NOTES

Chapter 6 : Mixed Assessment

ARGOPREP

1. What number is eight times as large as six?

A. 48

B. 64

C. 56

D. 72

SHOW YOUR WORK

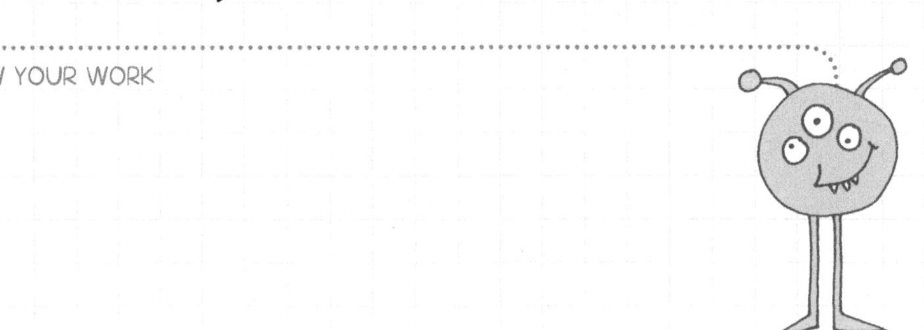

2. Thirty-two is eight times as large as what number?

A. 2

B. 3

C. 4

D. 5

SHOW YOUR WORK

3. What number is twice as large as nine?

A. 16

B. 18

C. 20

D. 22

SHOW YOUR WORK

4. Twenty-four is four times as large as what number?

A. 3

B. 4

C. 5

D. 6

SHOW YOUR WORK

5. Elizabeth plants trees on her farm. She plants three times as many trees as she planted last year. If she planted 10 trees last year, what problem represents the number of trees she planted this year?

A. 10 × 3

B. 10 ÷ 3

C. 10 + 3

D. 10 - 3

SHOW YOUR WORK

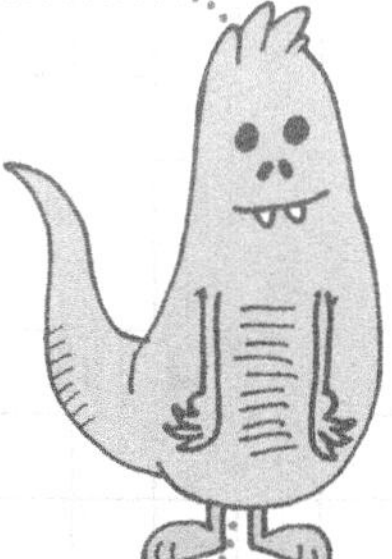

6. Based on the problem above, how many trees did Elizabeth plant this year?

A. 9

B. 13

C. 30

D. 31

SHOW YOUR WORK

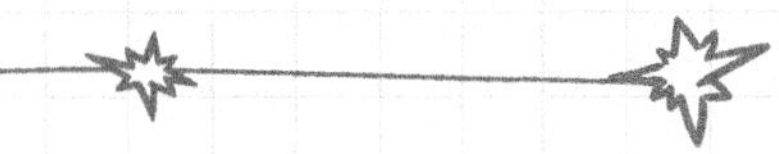

7. Joe buys an ice cream every week at school. If the ice cream costs **$2** and he has **$14**, which problem represents how many weeks he can buy ice cream?

A. $14 \times 2 = n$

B. $14 \div 2 = n$

C. $14 + 2 = n$

D. $14 - 2 = n$

SHOW YOUR WORK

8. Based on the problem above, how many weeks can Joe buy ice cream for?

A. 5 weeks

B. 7 weeks

C. 9 weeks

D. 12 weeks

SHOW YOUR WORK

9. What are two factors of 64?

A. 3, 8

B. 1, 7

C. 1, 8

D. 3, 7

SHOW YOUR WORK

10. 5 and 8 are factors of which number?

A. 40

B. 20

C. 32

D. 85

SHOW YOUR WORK

11. Which factor of 27 is prime?

A. 4

B. 1

C. 9

D. 3

SHOW YOUR WORK

12. Which factor of 88 is prime?

A. 22

B. 4

C. 8

D. 11

SHOW YOUR WORK

13. Which factor of 50 is composite?

A. 1

B. 5

C. 10

D. 2

SHOW YOUR WORK

14. Which factor of 42 is composite?

A. 2

B. 6

C. 7

D. 3

SHOW YOUR WORK

15. What answer shows multiples of 5?

A. 5, 10, 15

B. 5, 7, 9

C. 8, 10,12

D. 10, 20, 31

SHOW YOUR WORK

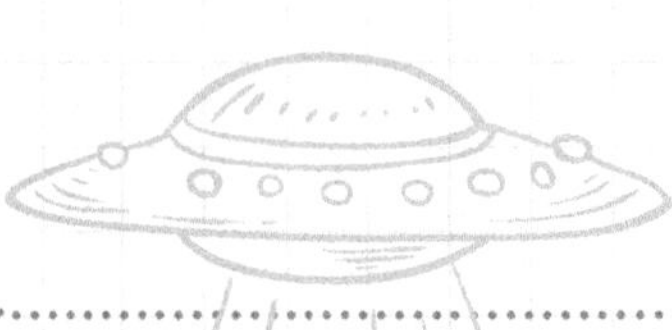

16. What is the next term in the pattern? Explain your reasoning.

3, 5, 8, 12, ______ ?

SHOW YOUR WORK

17. What is the rule of this pattern? Explain your reasoning.

35, 30, 60, 55, 110, 105, 210

SHOW YOUR WORK

18. Draw the next shape in the pattern. Explain your reasoning.

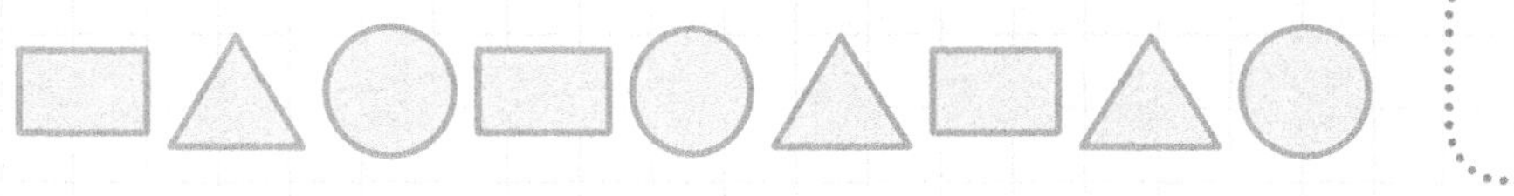

SHOW YOUR WORK

19. What is the missing number in the pattern? Explain your reasoning.

100 50 150 100 300 ______ 750 700

SHOW YOUR WORK

20. What is the rule of this pattern? Explain your reasoning.

99 90 82 75 66 58 51

SHOW YOUR WORK

NOTES

21. How does 560,000 compare to 5,600,000?

A. Ten times larger

B. Ten times smaller

C. Ten more

D. Ten less

SHOW YOUR WORK

22. How does 33,000 compare to 3,300?

A. Ten more

B. Ten less

C. Ten times smaller

D. Ten times larger

SHOW YOUR WORK

23. Which number is largest?

A. 600 + 70 + 1

B. 5000 + 60 + 7

C. 5000 + 600 + 70 + 1

D. 500 + 60 + 7

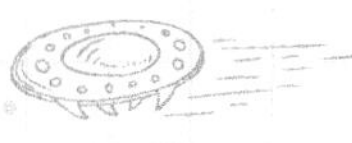

24. Which number is smallest?

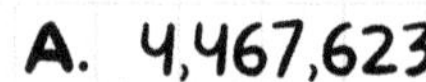

A. 4,467,623

B. 5,462,371

C. 40,572,281

D. 4,520,293

SHOW YOUR WORK

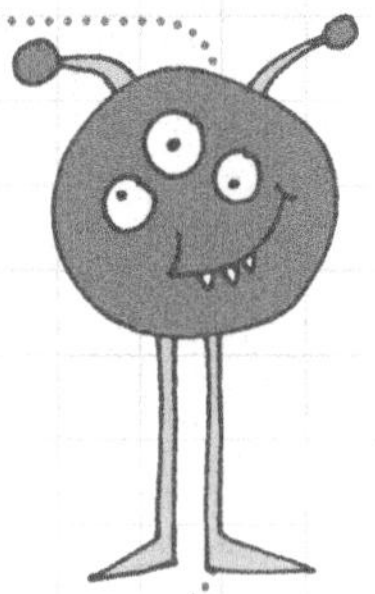

25. Which number is largest?

A. Ten thousand, six hundred eighty one

B. Ten thousand, six hundred

C. Ten thousand, eighty one

D. One thousand, six hundred eighty one

SHOW YOUR WORK

26. Which sign completes the comparison below?

3,478 ______ 3,748

A. >

B. <

C. ≥

D. ≤

SHOW YOUR WORK

27. Which sign completes the comparison below?

10,000+ 4,000+ 800+9 ______ 4,000 + 100 + 8

A. >

B. <

C. ≥

D. ≤

SHOW YOUR WORK

28. Which sign completes the comparison below?

32,873 ______ Thirty-two thousand seven hundred eighty three

A. >

B. <

C. ≥

D. ≤

SHOW YOUR WORK

29. Round to the nearest thousand. **6,482**

A. 6,500

B. 6,400

C. 7,000

D. 6,000

SHOW YOUR WORK

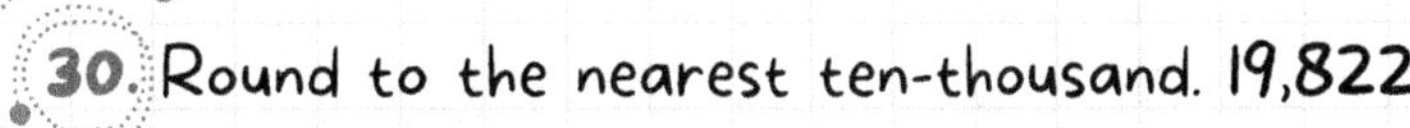

30. Round to the nearest ten-thousand. 19,822

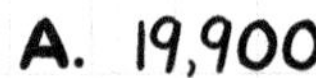

A. 19,900

B. 19,800

C. 20,000

D. 19,000

SHOW YOUR WORK

31. 2327 + 2836 =

A. 5612

B. 5613

C. 5163

D. 5162

SHOW YOUR WORK

32. 2348 - 456 =

A. 1892

B. 1982

C. 1800

D. 892

SHOW YOUR WORK

33. 27 x 6

A. 162

B. 126

C. 222

D. 120

SHOW YOUR WORK

34. 364 ÷ 7 =

A. 50

B. 52

C. 62

D. 71

SHOW YOUR WORK

35. 13,972 + 7,283 =

A. 22,155

B. 21,255

C. 25,215

D. 22,125

SHOW YOUR WORK

36. 42568 - 423 =

A. 42,415

B. 45,215

C. 42,145

D. 41,245

SHOW YOUR WORK

37. Multiply 632 x 8. Show your work.

SHOW YOUR WORK

38. Divide 268 ÷ 5. Show your work.

SHOW YOUR WORK

39. Samantha is saving money for a new bike. If she earns around **$5** a week in allowance and a new bike costs **$87**, about how long will it take her to save for a new bike? Explain your reasoning.

SHOW YOUR WORK

40. William wants to buy a computer. If the computer he wants costs **$545** and he has 15 weeks to save the money about how much money does he need to earn every week? Explain your reasoning.

SHOW YOUR WORK

NOTES

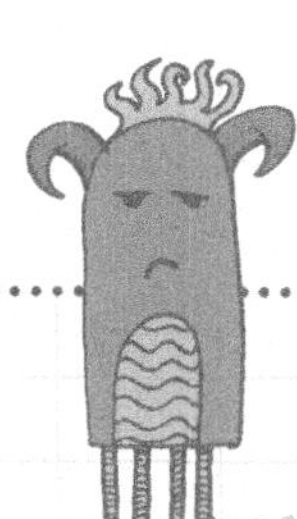

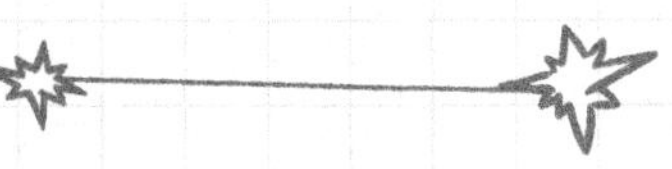

41. Which number could replace n to provide an equivalent fraction?

$\frac{4}{n} = \frac{24}{42}$

A. 7

B. 8

C. 9

D. 10

SHOW YOUR WORK

42. When comparing $\frac{3}{5}$ and $\frac{6}{10}$, which common denominator is easiest to use?

A. 50

B. 5

C. 20

D. 10

SHOW YOUR WORK

43. $\frac{7}{15}$ ______ $\frac{14}{30}$

A. >

B. <

C. =

D. ≥

SHOW YOUR WORK

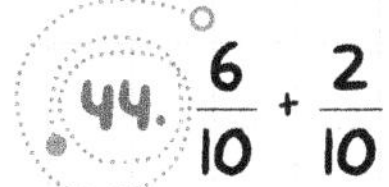

44. $\frac{6}{10} + \frac{2}{10} =$

A. $\frac{4}{10}$

B. $\frac{8}{10}$

C. $\frac{10}{10}$

D. $\frac{8}{20}$

SHOW YOUR WORK

45. Which fraction problem is equivalent to $\frac{13}{15}$?

A. $\frac{12}{15} + \frac{1}{15}$

B. $\frac{13}{1} + \frac{15}{1}$

C. $\frac{13}{1} + \frac{1}{15}$

D. $\frac{13}{15} + \frac{1}{1}$

SHOW YOUR WORK

46. Which fraction is not equivalent to $1\frac{2}{3}$?

A. $\frac{3}{3} + \frac{2}{3}$

B. $\frac{2}{3} + \frac{1}{3} + \frac{2}{3}$

C. $\frac{1}{3} + \frac{1}{3} + \frac{3}{3}$

D. $\frac{1}{3} + \frac{2}{3}$

SHOW YOUR WORK

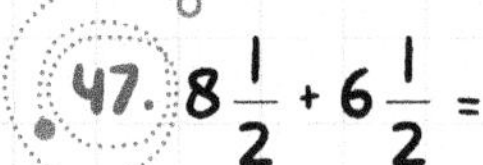

47. $8\frac{1}{2} + 6\frac{1}{2} =$

A. $14\frac{1}{2}$

B. 15

C. $15\frac{1}{2}$

D. 16

SHOW YOUR WORK

48. $5\frac{1}{6} - 2\frac{4}{6} =$

A. $2\frac{3}{6}$

B. $3\frac{3}{6}$

C. $3\frac{2}{6}$

D. $2\frac{2}{6}$

SHOW YOUR WORK

49. Jacey has a goal to work out 6 hours a week. She rides her bike for $\frac{1}{2}$ hour for three days and walks with her mom for 1 hour another day. How many more hours does she need to exercise to reach her goal?

A. $2\frac{1}{2}$ hours

B. 3 hours

C. $3\frac{1}{2}$ hours

D. 4 hours

SHOW YOUR WORK

6 | Mixed Assessment

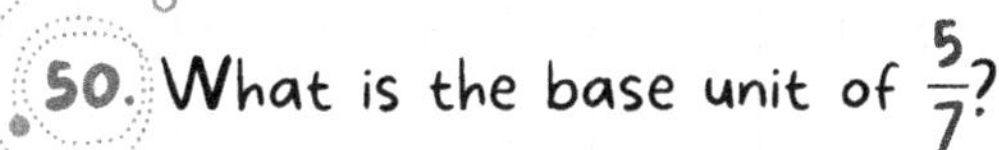

50. What is the base unit of $\frac{5}{7}$?

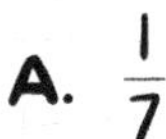

A. $\frac{1}{7}$

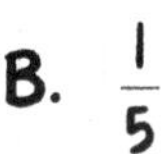

B. $\frac{1}{5}$

C. $\frac{7}{5}$

D. $\frac{3}{7}$

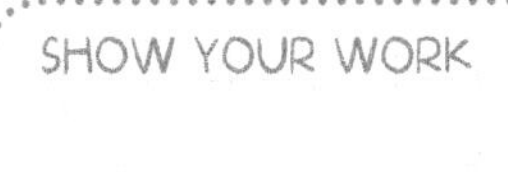

SHOW YOUR WORK

51. What is $8 \times \frac{1}{3}$?

A. $\frac{3}{8}$

B. $8\frac{1}{3}$

C. $1\frac{3}{8}$

D. $\frac{8}{3}$

SHOW YOUR WORK

52. Ryan is helping his mother bake a bunch of cupcakes for a school fundraiser. For one batch of cupcakes, they use $\frac{1}{4}$ of a dozen eggs. They want to make **6** batches of cupcakes. What problem represents how many eggs they need per batch of cupcakes?

A. $\frac{1}{4} \times 6$

B. $\frac{1}{4} \times 12$

C. $\frac{1}{4} + 12$

D. $12 - \frac{1}{4}$

SHOW YOUR WORK

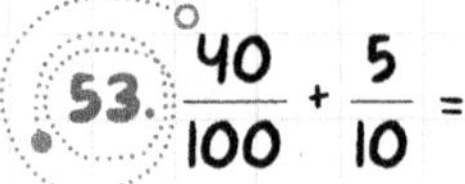

53. $\frac{40}{100} + \frac{5}{10} =$

A. $\frac{45}{100}$

B. $\frac{54}{100}$

C. $\frac{90}{100}$

D. $\frac{45}{100}$

SHOW YOUR WORK

54. $\frac{2}{10} + \frac{75}{100}$

A. $\frac{95}{100}$

B. $\frac{77}{100}$

C. $\frac{77}{110}$

D. $\frac{90}{100}$

SHOW YOUR WORK

55. Write the equivalent fraction to 0.53

A. $\frac{1}{53}$

B. $\frac{53}{100}$

C. $\frac{53}{10}$

D. $\frac{5}{10}$

SHOW YOUR WORK

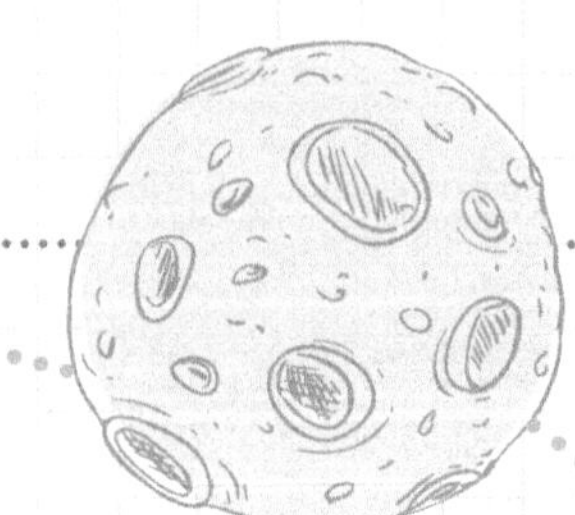

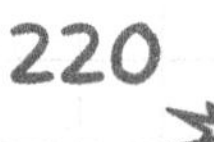

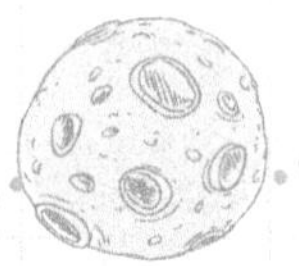

56. A pie was divided into 16 pieces. We served it to five friends and 3 family members. Write a problem that would represent how much pie was left. Explain your reasoning.

SHOW YOUR WORK

57. Write the equivalent decimal to $\frac{27}{100}$. Show your work.

SHOW YOUR WORK

58. Write a number smaller than 0.6. Explain your reasoning.

SHOW YOUR WORK

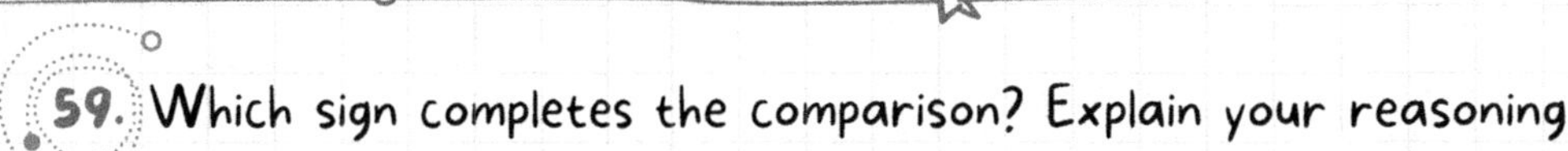

59. Which sign completes the comparison? Explain your reasoning.

0.86 ________ 0.68

SHOW YOUR WORK

60. Write the equivalent fraction to 0.76. Show your work.

SHOW YOUR WORK

NOTES

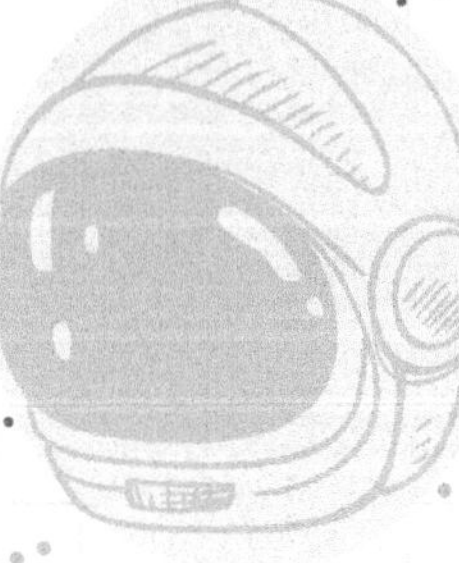

61. Which measurement completes the table below?

Centimeters	Meters
86	x

A. 0.86

B. 8.6

C. 86

D. 860

SHOW YOUR WORK

62. Which measurement is equal to 8 hours?

A. 48 minutes

B. 480 minutes

C. 4.8 minutes

D. 64 minutes

SHOW YOUR WORK

63. Which measurement is equal to 5,612 milliliters?

A. 5.612 L

B. 56.12 L

C. 561.2 L

D. 5,612,000 L

SHOW YOUR WORK

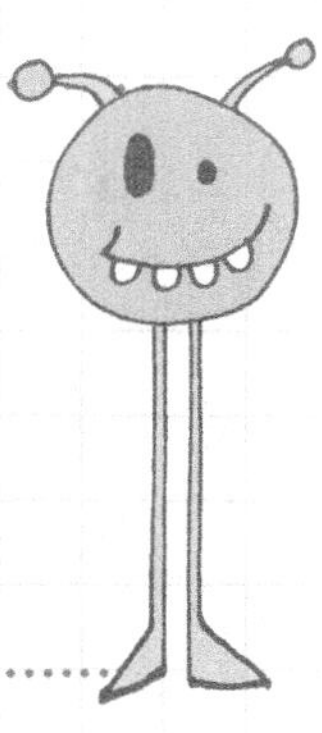

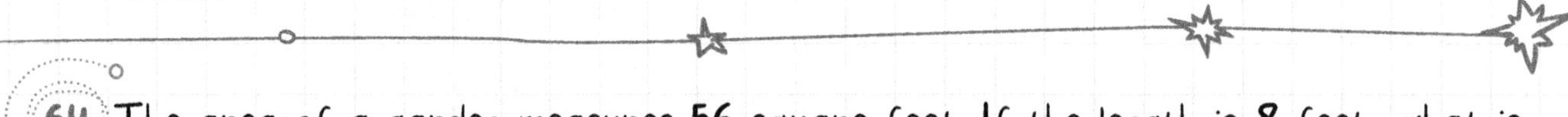

64. The area of a garden measures 56 square feet. If the length is 8 feet, what is the width?

A. 5 feet

B. 6 feet

C. 7 feet

D. 8 feet

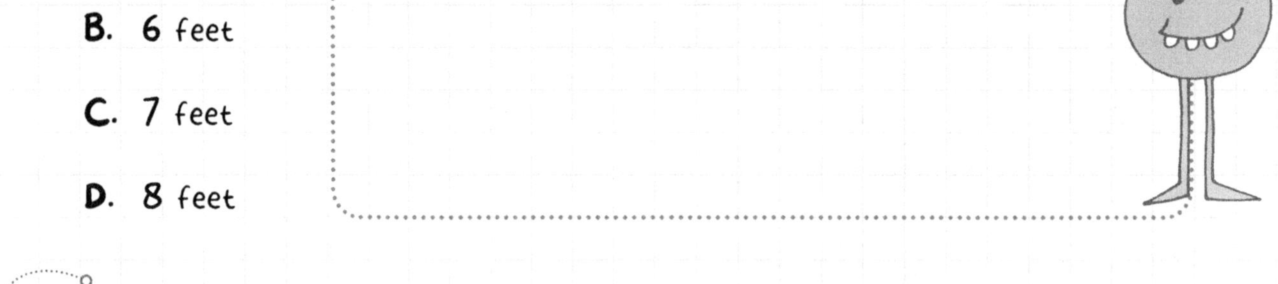

65. The perimeter of a garden measures 66 feet. If the width is 11 feet, what is the length?

A. 22 ft

B. 11 ft

C. 44 feet

D. 10 feet

66. If the rectangle has an area of 75 and a width of 5, what is the perimeter?

A. 30

B. 15

C. 20

D. 40

67. Which number could be placed on the line plot below?

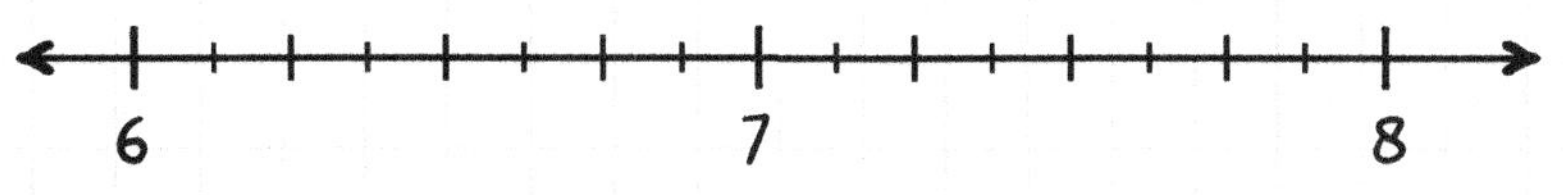

A. $5\frac{1}{4}$

B. $8\frac{2}{8}$

C. $7\frac{3}{2}$

D. $7\frac{3}{4}$

SHOW YOUR WORK

68. Which angle is the largest?

A. $56\frac{1}{2}°$

B. $51\frac{2}{6}°$

C. $53\frac{1}{2}°$

D. $56\frac{2}{3}°$

SHOW YOUR WORK

69. Which angle is the smallest?

A. $56\frac{1}{2}°$

B. $51\frac{2}{3}°$

C. $53\frac{1}{2}°$

D. $56\frac{2}{3}°$

SHOW YOUR WORK

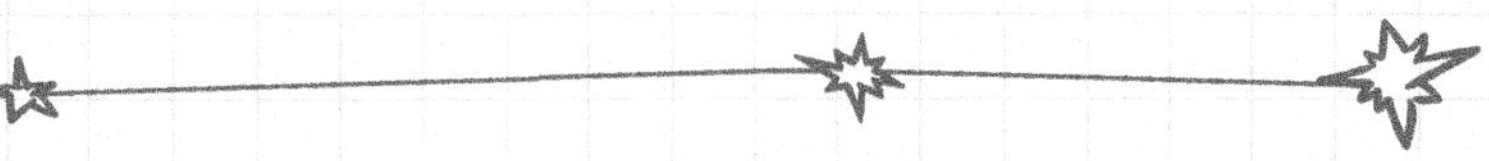

Use the drawing below to answer questions 70-72.

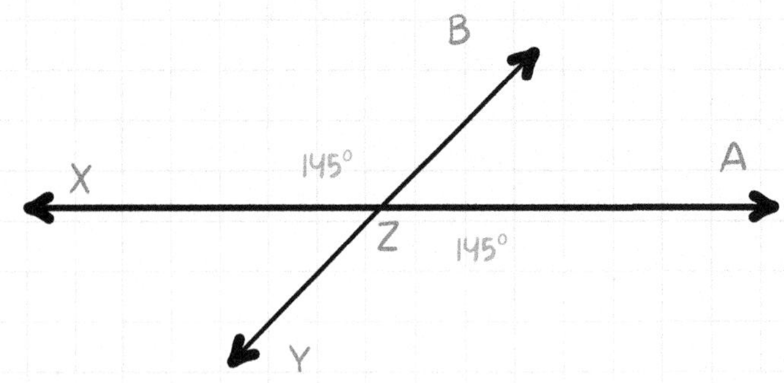

70. What is the measure of angle YZA?

A. 35°

B. 145°

C. 180°

D. 90°

SHOW YOUR WORK

71. Which angle measures 145°?

A. XZY

B. AZX

C. XZB

D. ZAY

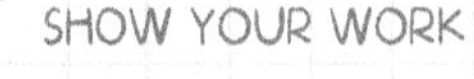

SHOW YOUR WORK

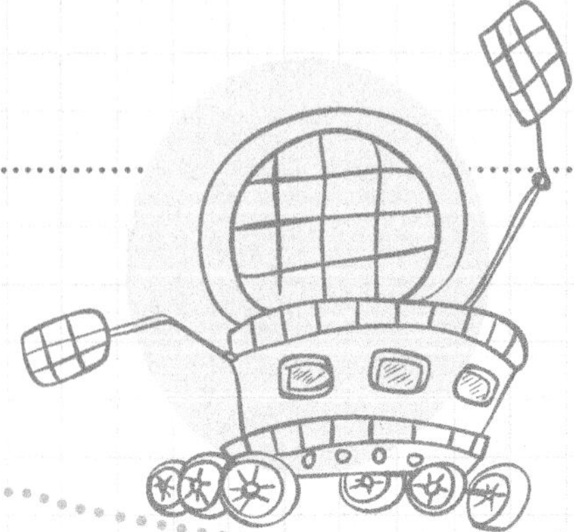

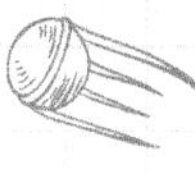

72. What is the measure of angle XZA?

A. 180°

B. 145°

C. 35°

D. 90°

SHOW YOUR WORK

73. If angle x is 130° and it is divided into 2 equal angles, what is the measure of one of the angles?

A. 55°

B. 65°

C. 56°

D. 75°

SHOW YOUR WORK

74. If the clock is set at 4:15 and the minute hand moves 90°, what is the new time?

A. 4 : 30

B. 4 : 45

C. 5 : 15

D. 5 : 30

SHOW YOUR WORK

75. If the clock is set at 11:00 and the minute hand moves 180°, what is the new time?

A. 11 : 15

B. 11 : 30

C. 11 : 45

D. 12:00

SHOW YOUR WORK

76. If the clock is set at 7 : 30 and the minute hand moves 360°, what is the new time?

A. 8 : 30

B. 7 : 45

C. 8 : 45

D. 8 : 00

SHOW YOUR WORK

77. Draw an angle that is larger than 90° and smaller than 180°. Explain how you completed your drawing.

SHOW YOUR WORK

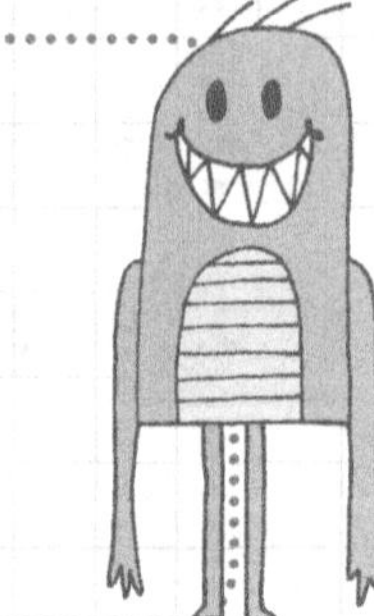

78. Use a protractor to measure the following angle.
Explain how you completed the measurement.

SHOW YOUR WORK

79. Use a protractor to draw an angle that measures around 110.
Explain how you completed your drawing.

SHOW YOUR WORK

80. Use a protractor to measure the following angle.
Explain how you completed the measurement.

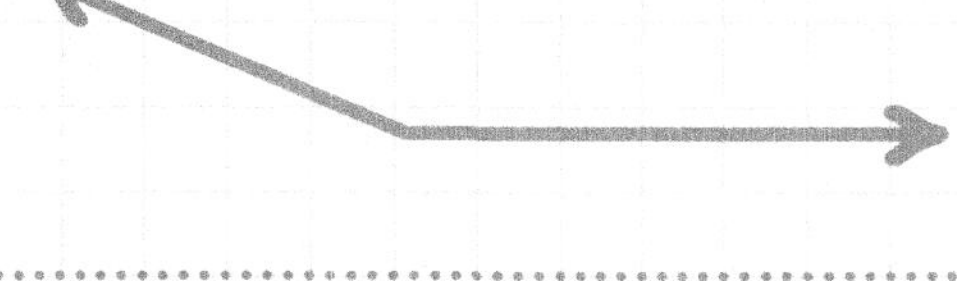

SHOW YOUR WORK

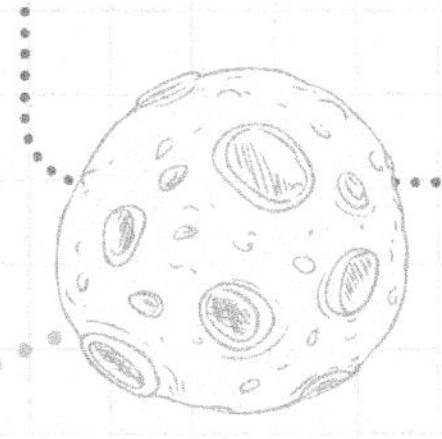

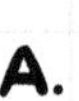

81. Identify the line.

A.

C.

B.

D.

SHOW YOUR WORK

82. Identify the ray.

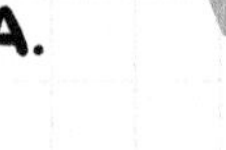

A.

C.

B.

D.

SHOW YOUR WORK

83. Which angle is acute?

A. 

C. 

B. 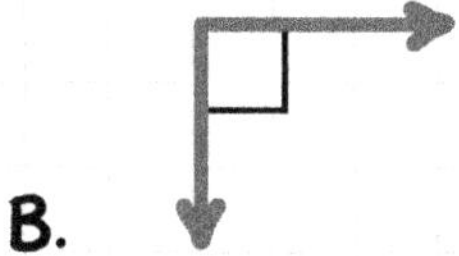

D.

SHOW YOUR WORK

84. Which angle is obtuse?

A.

C. 

B.

D. 

SHOW YOUR WORK

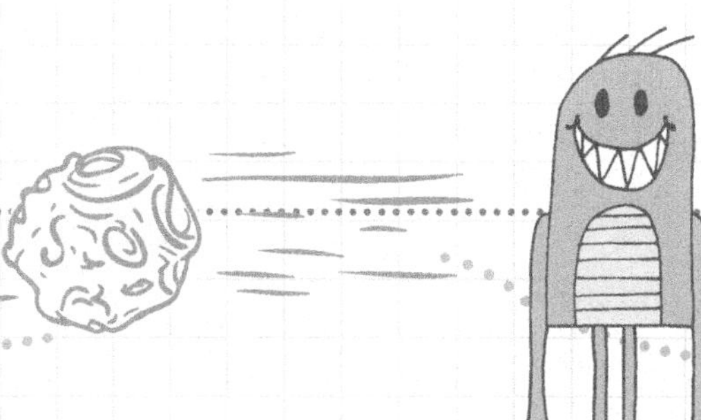

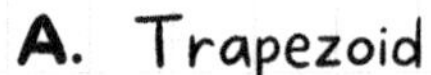

85. Which shape contains 1 set of parallel lines?

A. Trapezoid

B. Triangle

C. Circle

D. Pentagon

SHOW YOUR WORK

86. Which shape does not contain 2 sets of perpendicular lines?

A. Rectangle

B. Square

C. Diamond

D. Triangle

SHOW YOUR WORK

87. Which shape does not contain parallel or perpendicular lines?

A. Trapezoid

B. Circle

C. Rectangle

D. Rhombus

SHOW YOUR WORK

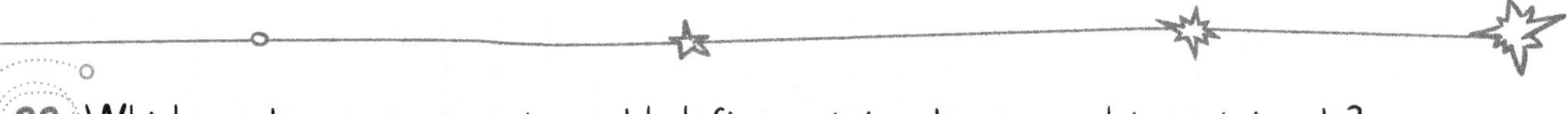

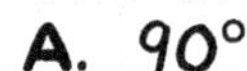

88. Which angle measurement would define a triangle as an obtuse triangle?

A. 90°

B. 24°

C. 110°

D. 72°

89. What type of triangle contains the following angle measurements? 45, 55, 80

A. Equilateral

B. Right

C. Isosceles

D. Scalene

SHOW YOUR WORK

90. What type of triangle contains the following angle measurements? 60, 60, 60

A. Equilateral

B. Right

C. Isosceles

D. Scalene

SHOW YOUR WORK

91. What type of triangle contains the following angle measurements? 90°

A. Equilateral

B. Right

C. Isosceles

D. Scalene

SHOW YOUR WORK

92. How do you know a triangle is isosceles?

A. Three angles under 90°

B. An angle larger than 90°

C. 2 equal angles

D. 3 equal angles

93. Which triangle does not make sense?

A. 100°, 40°, 40°

B. 120°, 30°, 40°

C. 40°, 60°, 80°

D. 60°, 60°, 60°

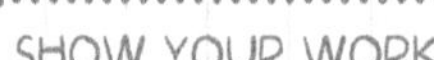

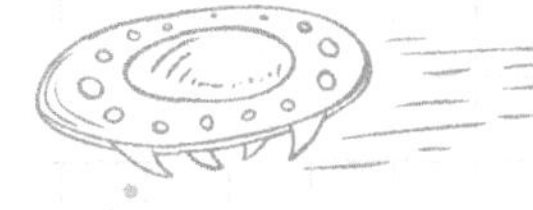

94. Which line of symmetry is correct?

A. 

B.

C. 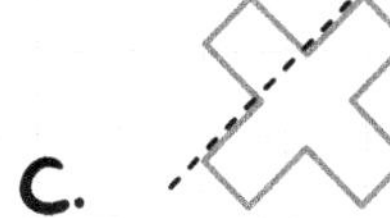

D. 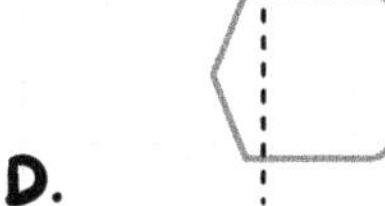

SHOW YOUR WORK

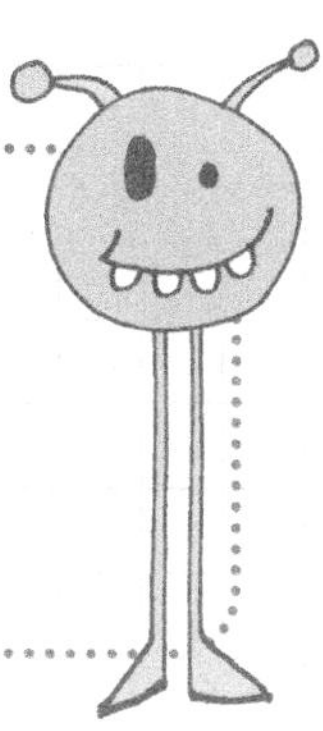

95. Which shape does not have a line of symmetry?

A. 

B. 

C. 

D. 

SHOW YOUR WORK

96. Draw a shape that does not include perpendicular or parallel lines. Explain how you completed your drawing.

SHOW YOUR WORK

97. Draw a picture of a right triangle. Explain how you completed your drawing.

SHOW YOUR WORK

98. Draw shape with one line of symmetry. Explain how you completed your drawing.

SHOW YOUR WORK

99. Draw the lines of symmetry in the following shape. Explain your reasoning.

SHOW YOUR WORK

100. Draw a shape with no lines of symmetry. Explain your reasoning.

SHOW YOUR WORK

ANSWER SHEET

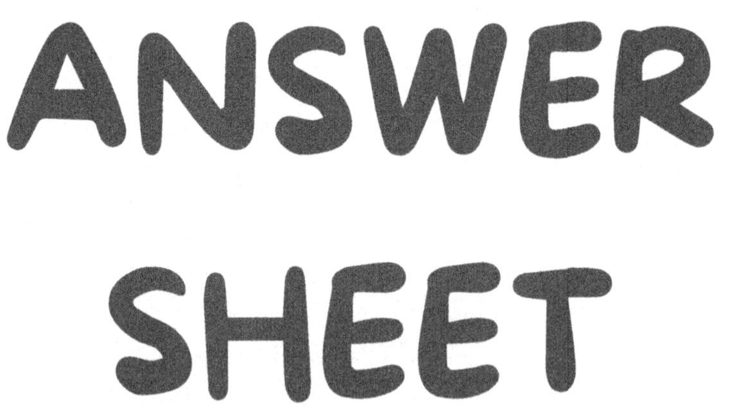

ANSWER SHEET

Chapter 1: Operations & Algebraic Thinking

1.1.A Use the four operations with whole numbers to solve problems.

1. Answer: A. 24
2. Answer: 8
3. Answer: D. 5
4. Answer: 7, 3 (order does not matter)
5. Answer: B. 4
6. Answer: C. 12 x 4 = ?
7. Answer: D. 58 candy bars
8. Answer: C. (15 x 2) + 15 = ?
9. Answer: D. 45 stamps
10. Answer: $6.00

1.1.B Use the four operations with whole numbers to solve problems.

1. Answer: A. 7 x 3 = ?
2. Answer: B. 3 x ? =21
3. Answer: C. 12 x 6 =?
4. Answer: 7 x ? = 63
5. Answer D. 2 x ? = 96
6. Answer: A. $63 ÷ 9 = $7
7. Answer: B. $36 ÷ 6 = 6 packs
8. Answer: C. 320 ÷ 8 = 40 cookies
9. Answer: B. 96 ÷ 3 = 32
10. Answer: 8 + 9 = 17
    17 x 7 = 119
    Jane has 119 candies and cupcakes altogether.

1.1.C Use the four operations with whole numbers to solve problems.

1. Answer: A. 465 - 375, to determine how much was paid by the students in the club
2. Answer: You divide the answer by 5.
3. Answer: B. 4 x ? = 300

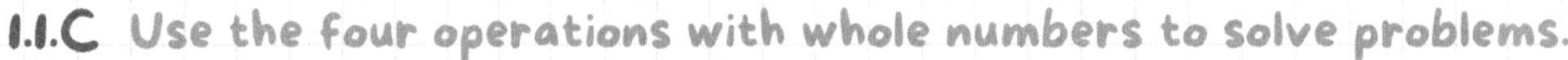

4. Answer: D. \$19 x 8 = \$152
5. Answer: He should buy more boxes than he needs so he has at least 87 markers.
6. Answer: D. \$50 - \$5 - \$17 - \$17 = \$11
7. Answer: 75 x 3 = 225 total pieces of candy
   225 ÷ 5 = 45 pieces of candy per person
   Each person will receive 45 pieces of candy.
8. Answer: \$10 x 7 = \$70 from delivering pizzas
   \$11 x 3 = \$33 from babysitting
   \$70 + \$33 = \$103
   Gabe earned \$103 last week.
9. Answer: D. Tyler slept for 4 hours and if Will slept 5 hours longer than Tyler, then (4 + 5 = 9). Therefore, Will slept for 9 hours.
10. Answer: C. 15 x 32 = 480
    480 ÷ 7 = 68 r 4

Since the remainder is 4, that means there will be 4 beads left over.

1.2. Gain familiarity with factors and multiples.

1. Answer: A. 8 x 8 = 64
2. Answer: 1, 2, 4, 5, 10, 20
3. Answer: B. 3 cannot be broken into smaller factors.
4. Answer: C. 18 is a result of 2 x 9.
5. Answer: D. 9 times 6 gives you 54.
6. Answer: B. 8 times 4 gives you 32.
7. Answer: D. 5 cannot be divided into smaller factors.
8. Answer: It can be broken into the two smaller factors of 2 and 3.
9. Answer: D. You can break 21 into two smaller factors: 3 and 7.
10. Answer A. 7 times 11 is 77.

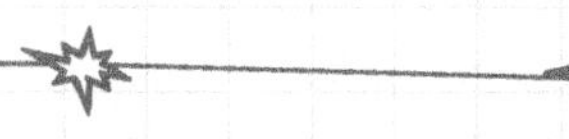

ANSWER SHEET

1.3. Generate and analyze patterns.

1. Answer B. Each term is 2 less that the term before it.
2. Answer A. The pattern is a repetition of triangle then square.
3. Answer C. Each number is double the number before it.
4. Answer: One possible response is 100, 95, 90, 85, 80.
5. Answer: One possible answer is below

   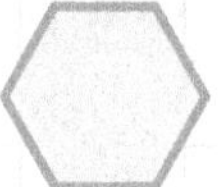

6. Answer C. The rule is -3 and 4-3 is 1.
7. Answer B. This pattern is two shapes repeated and the next shape to be repeated is a square.
8. Answer A. The pattern is -10 so the next term is 70.
9. Answer B. The pattern is x3 so the next term is 9 x 3.
10. Answer C. The next term is a lightning rod, as the pattern is plus sign, heart, lightning rod.

1.4. Chapter Test

1. Correct answer is A because 7 times 5 is 35.
2. Correct answer is B because six times eight is forty-eight.
3. Correct answer is A because we know Alicia collects three times as many as Sara and Sara collected 21 flowers. Therefore, 21 times 3 is our answer.
4. Correct answer is C because 6 x 2 = 12.
5. The correct answer is D because she has eight weeks with three dollars each week.
6. The correct answer is A because you need to figure out what number will get you closest to $47.
7. The correct answer is C. He will need at least 10 weeks of allowance to get him enough money to buy his father's present.
8. Answer: To buy something, you need more money not less, so you need to round up to the next whole number.
9. The correct answer is D. 3 x 18 = 54
10. Correct answer is D. 3 is a prime number because it cannot be broken into smaller factors.

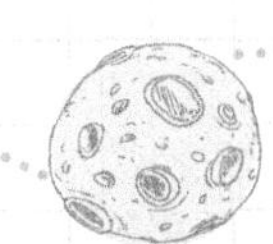

11. The correct answer is C. 4 can be broken down into the factors of 2 x 2.
12. The correct answer is B. 3 x 2= 6, 3 x 3 =9, 3 x 4 = 12
13. The correct answer is C. Each number is the previous number multiplied by 10.
14. Correct answer is down arrow. The pattern is right, down, right, left, up, right, down, right, left, up.
15. The correct answer is 30. The rule for this pattern is + 2, then x 2.

Chapter 2: Numbers and Operation in Base Ten

2.1.A Generalize place value understanding for multi-digit whole numbers.

1. The correct answer is C. Multiplying 480 by 10 increases its value to 4,800.
2. The correct answer is A. Dividing 20 by 10 decreases its value to 2.
3. The correct answer is B. Multiplying 36 by 10 increases its value to 360.
4. Correct answer: D. 11,000 is 10 times larger than 1, 100.
5. Answer: Multiply it by 10.
6. The correct answer is C. Multiplying ten by 380= 3,800.
7. The correct answer is D. Dividing 240 by 10= 24.
8. Answer: It is ten times larger.
9. Correct answer is A. 960 is ten times smaller than 9,600.
10. Answer: Divide by 10.

2.1.B Generalize place value understanding for multi-digit whole numbers.

1. Answer: 3, 807
2. The correct answer is A.
3. The correct answer is C.
4. The correct answer is B.
5. The correct answer is D.
6. The correct answer is 792, 632 because 90,000 is larger than 20,000.
7. The correct answer is D. 3,582 is smaller than 3, 852.
8. The correct answer is B. 78,968 is smaller than 87,689.
9. The correct answer is A. 50,000 is larger than 5,000.
10. The correct answer is A. 400,000 is larger than 40,000

ANSWER SHEET

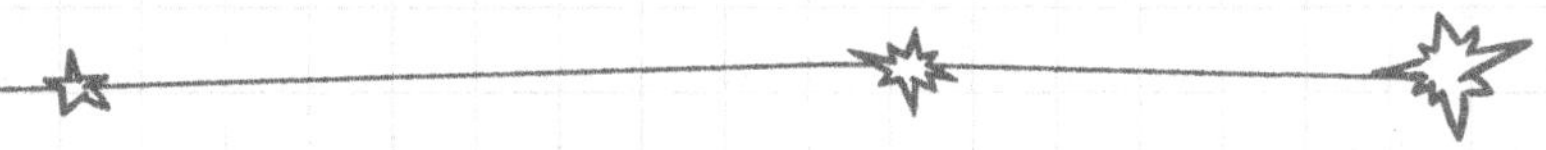

2.1.C Generalize place value understanding for multi-digit whole numbers.

1. Answer: 735,000 because 700 rounds to 5,000.
2. The correct answer is A. It remains 9,000 because 30 is smaller than 50.
3. The correct answer is B. 69 rounds up to 70 because 8 is larger than 5.
4. The correct answer is D. 312 rounds down to 300 as 1 is smaller than 5.
5. The correct answer is C. 68 rounds to 70 as 8 is larger than 5.
6. The correct answer is 700. 34 keeps 700 the same as it is smaller than 50.
7. The correct answer is A. 80 rounds 900 up to 1,000.
8. The correct answer is B. This is the only choice that would correctly round to 1,100.
9. The correct answer is C. This is the only choice that would correctly round to 1200.
10. The correct answer is D. This is the only choice that would round to $120.

2.2.A Use place value understanding and properties of operations to perform multi-digit arithmetic.

1. The correct answer is B.
2. The correct answer is D.
3. The correct answer is C.
4. The correct answer is A.
5. The correct answer is D.
6. The correct answer is C.
7. The correct answer is B.
8. The correct answer is B.
9. The correct answer is A.
10. The correct answer is D.

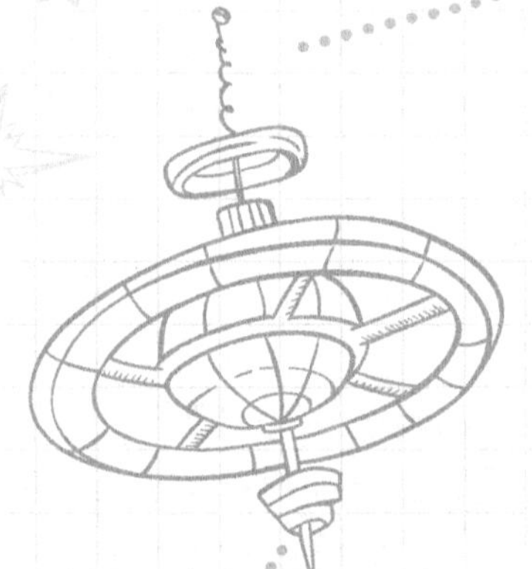

2.2.B Use place value understanding and properties of operations to

1. The correct answer is C.
2. The correct answer is A.
3. The correct answer is B.
4. The correct answer is D.
5. The correct answer is B.

ANSWER SHEET

6. The correct answer is C.
7. The correct answer is A.
8. The correct answer is D.
9. The correct answer is A.
10. The correct answer is B.

2.2.C Use place value understanding and properties of operations to perform multi-digit arithmetic

1. The correct answer is D.
2. The correct answer is A.
3. The correct answer is B.
4. The correct answer is C.
5. The correct answer is B.
6. The correct answer is A.
7. The correct answer is B.
8. The correct answer is C.
9. The correct answer is D.
10. The correct answer is A.

2.3. Chapter Test

1. The correct answer is A. 34,000 is ten times larger than 3,400.
2. The correct answer is B. 78,000 is ten times larger than 7,800.
3. The correct answer is C. 5 in the ones column is larger than 2 in the ones column.
4. The correct answer is B. 79 is smaller than 95.
5. The correct answer is B. 809, 098 is smaller than 890, 809.
6. The correct answer is A. 1 rounds up to 2 because 6 is larger than 5.
7. The correct answer is B. That is the only choice that rounds to 2,000 correctly.
8. The correct answer is C. That is the only choice that rounds to 200,000 correctly.
9. The correct answer is D.
10. The correct answer is D.
11. The correct answer is A.
12. The correct answer is B.
13. The correct answer is C.
14. The correct answer is D.
15. The correct answer is A.

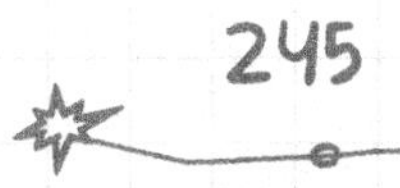

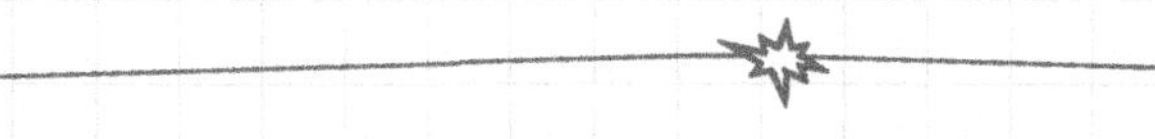

Chapter 3: Fractions

3.1. Extend understanding of fraction equivalence and ordering.

1. The correct answer is A. 4x3= 12 and 7x3= 21.
2. The correct answer is B. 3÷3= 1 and 9÷3= 3.
3. The correct answer is D. 1x4=4 and 6x4= 24.
4. The correct answer is C. 15÷3= 5 and 25÷5= 5.
5. The correct answer is A. All other choices contain multiples of 33 and 66.
6. The correct answer is A. 7x4= 28 so the correct answer contains 16÷4.
7. The correct answer is D. 6÷3= 2, so the correct answer contains 12 x 3.
8. The correct answer is C. 18÷3= 6 so the correct answer contains 12÷3.
9. The correct answer is B. 18÷2= 9 so the correct answer contains 81÷9.
10. One possible answer is:

1. The correct answer is B. Both fractions can be rewritten with a denominator of 24.
2. The correct answer is C. Both fractions can be rewritten with a denominator of 21.
3. The correct answer is A. The easiest comparison is to be made by keeping $\frac{2}{6}$ the same.
4. The correct answer is D. Both fractions can be rewritten with a denominator of 30.
5. The correct answer is C. Both fractions can be rewritten with a denominator of 24.
6. The correct answer is A. $\frac{2}{3}$ is larger than $\frac{6}{12}$.
7. The correct answer is B. $\frac{12}{18}$ is larger than $\frac{4}{7}$.
8. The correct answer is B. $\frac{4}{6}$ is larger than $\frac{1}{4}$.

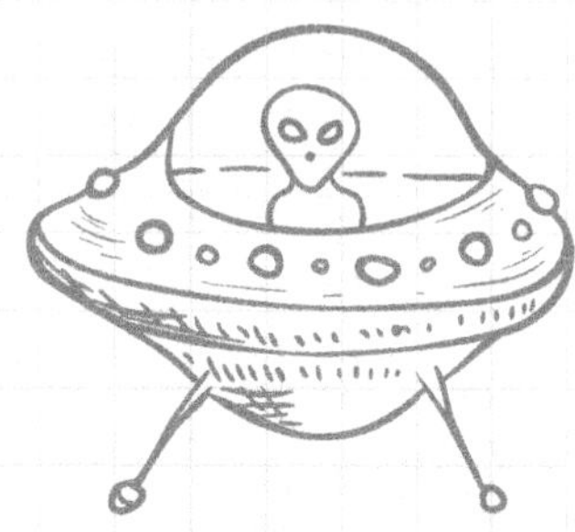

9. The correct answer is B. $\frac{8}{9}$ is larger than $\frac{7}{8}$.

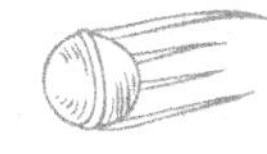

10. The correct answer is A. $\frac{3}{5}$ is larger than $\frac{2}{7}$

3.2.A Build fractions from unit fractions.

1. The correct answer is D. When adding fractions, you add the numerator and the denominator stays the same.
2. The correct answer is C. When adding fractions, you add the numerator and the denominator stays the same.
3. The correct answer is B. When adding fractions, you add the numerator and the denominator stays the same.
4. The correct answer is D. You can think of this problem as a fraction problem: $\frac{9}{9} - \frac{4}{9} =$
5. One possible answer would be: $\frac{8}{8} - \frac{6}{8} =$

3.2.B Build fractions from unit fractions.

1. The correct answer is A. You can break $\frac{7}{8}$ into $\frac{3}{8}$ and $\frac{4}{8}$
2. The correct answer is C. You can break $\frac{9}{12}$ into $\frac{3}{12}$, three times.
3. The correct answer is B. You can break $2\frac{3}{4}$ into $\frac{4}{4} + \frac{4}{4} + \frac{3}{4}$
4. The correct answer is D. $\frac{5}{6} + \frac{1}{6}$ does not equal $\frac{5}{6}$.
5. The correct answer is D. $\frac{4}{3} + \frac{3}{3}$ does not equal $3\frac{3}{4}$.

3.2.C Build fractions from unit fractions.

1. The correct answer is B. Remember to complete both the whole number and the fraction operation.
2. The correct answer is C. Remember to complete both the whole number and the fraction operation.
3. The correct answer is A. Remember to complete both the whole number and the fraction operation.

ANSWER SHEET

4. The correct answer is D. Remember, you need to borrow from the mixed number to subtract the fraction.
5. The correct answer is A. Remember to complete both the whole number and the fraction operation.

3.2.D Build fractions from unit fractions.

1. The correct answer is B. To find the answer you add $\frac{3}{4} + \frac{3}{4} + \frac{3}{4}$.
2. The correct answer is C. To find the answer you add $\frac{2}{8} + \frac{3}{8}$.
3. The correct answer is A. To find the answer, you take $\frac{8}{8} - \frac{3}{8}$.
4. The correct answer is D. To find the answer, you take $7\frac{3}{10} - 4\frac{6}{10}$
5. The correct answer is A. To find the answer, you take $12\frac{3}{10} + 6\frac{4}{10}$.

3.2.E Build fractions from unit fractions.

1. The correct answer is A. To find the answer, $\frac{6}{7}$ is 6 of the base unit $\frac{1}{7}$.
2. The correct answer is D. To find the answer, $\frac{3}{4}$ is 3 of the base unit $\frac{1}{4}$.
3. The correct answer is C. $\frac{8}{12}$ is 8 of the base unit $\frac{1}{12}$.
4. The correct answer is B. $\frac{13}{20}$ is 13 of the base unit $\frac{1}{20}$.
5. The correct answer is A. To find the answer, you multiply 1 x 7 over 10.

3.2.F Build fractions from unit fractions.

1. The correct answer is D. The problem becomes $21 \times \frac{1}{8}$.
2. The correct answer is C. The problem becomes $10 \times \frac{1}{6}$.

3. The correct answer is B. The problem becomes $44 \times \frac{1}{12}$.
4. The correct answer is A. The problem becomes $6 \times \frac{1}{9}$.
5. The correct answer is D. The problem becomes $12 \times \frac{1}{7}$.

3.2.G Build fractions from unit fractions.

1. The correct answer is B. $\frac{3}{4} \times 4 = 12 \times \frac{1}{4}$
2. The correct answer is D. $6 \times \frac{1}{4} = 24 \times \frac{1}{4}$
3. The correct answer is A. $\frac{3}{5} \times 2 = 6 \times \frac{1}{5}$
4. The correct answer is C. $\frac{2}{14} \times 3 = 6 \times \frac{1}{14}$
5. The correct answer is C. They have $\frac{1}{2} \times 4$ books or $\frac{4}{2}$ books, which simplifies to 2.

3.3.A Understand decimal notation for fractions, and compare decimal fractions.

1. The correct answer is B. The problem becomes $\frac{70}{100} + \frac{3}{100}$.
2. The correct answer is D. The problem becomes $\frac{20}{100} + \frac{62}{100}$.
3. The correct answer is A. The problem becomes $\frac{30}{100} + \frac{6}{100}$.
4. The correct answer is C. The problem becomes $\frac{90}{100} + \frac{1}{100}$.
5. The correct answer is A. The problem becomes $\frac{40}{100} + \frac{24}{100}$.
6. The correct answer is B. The problem becomes $\frac{10}{100} + \frac{7}{100}$.

ANSWER SHEET

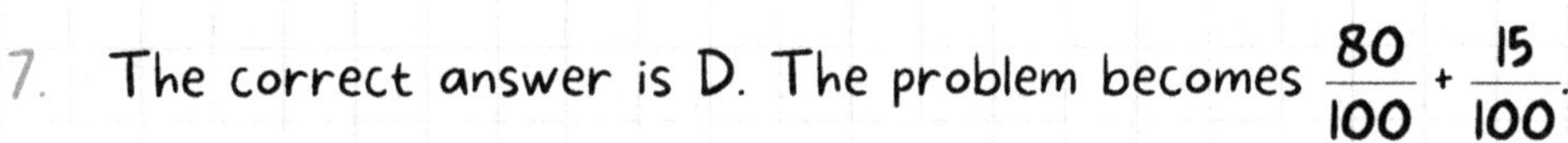

7. The correct answer is D. The problem becomes $\frac{80}{100} + \frac{15}{100}$.

8. The correct answer is C. The problem becomes $\frac{60}{100} + \frac{33}{100}$.

9. The correct answer is A. The problem becomes $\frac{80}{100} + \frac{4}{100}$.

10. The correct answer is D. The problem becomes $\frac{90}{100} + \frac{2}{100}$.

3.3.B Understand decimal notation for fractions, and compare decimal fractions.

1. The correct answer is D.
2. The correct answer is A.
3. The correct answer is B.
4. The correct answer is C.
5. The correct answer is A.
6. The correct answer is A.
7. The correct answer is D.
8. The correct answer is C.
9. The correct answer is A.
10. The correct answer is B.

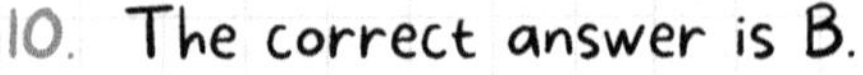

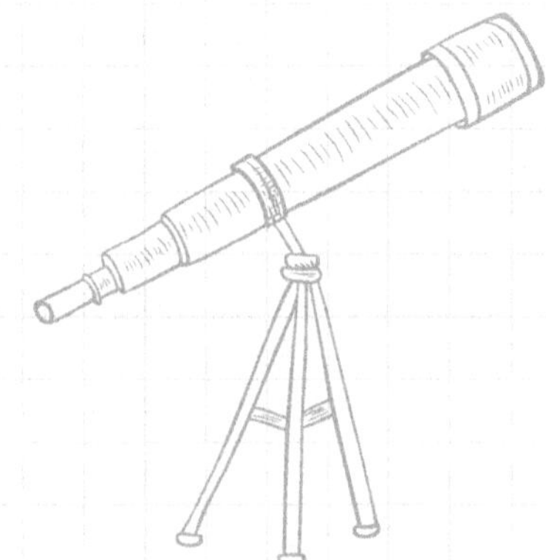

3.3.C Understand decimal notation for fractions, and compare decimal fractions.

1. The correct answer is A. $\frac{80}{100}$ is larger than any other number.

2. The correct answer is D. $\frac{17}{100}$ is smaller than any other number.

3. The correct answer is C. $\frac{67}{100}$ is larger than any other number.

4. The correct answer is A. $\frac{4}{10}$ is smaller than any other number.

5. The correct answer is B. $\frac{79}{100}$ is larger than any other number.

ANSWER SHEET

6. The correct answer is A. $\frac{65}{100}$ is larger than $\frac{36}{100}$.
7. The correct answer is B. $\frac{50}{100}$ is smaller than $\frac{74}{100}$.
8. The correct answer is B. $\frac{42}{100}$ is larger than $\frac{80}{100}$.
9. The correct answer is A. $\frac{63}{100}$ is larger than $\frac{23}{100}$.
10. The correct answer is A. $\frac{90}{100}$ is older than $\frac{45}{100}$.

3.4. Chapter Test

1. The correct answer is B. 12 x 3 = 36 and 2 x 3 = 6.
2. The correct answer is C. 6 x 3 = 18, and 18 x 1 = 18
3. The correct answer is B. When converted to like denominators, $\frac{4}{5} = \frac{32}{40}$ and $\frac{7}{8} = \frac{35}{40}$
4. The correct answer is A. When adding 2 fractions, the denominators stay the same.
5. One possible answer would be: $\frac{2}{12} + \frac{2}{12} + \frac{1}{12} + \frac{1}{12}$
6. The correct answer is D. $\frac{12}{18}$ can be broken into $\frac{6}{18} + \frac{6}{18}$.
7. The correct answer is B. Your goal should be to have a value of 7 in the numerator.
8. The correct answer is C. Remember to combine both the mixed number and fraction.
9. The correct answer is C. Remember, you need to borrow from the mixed number to subtract the fraction.
10. The correct answer is A. $\frac{1}{2} + \frac{1}{2} + \frac{1}{2} + \frac{1}{2}$ = 2 hours 3 - 2 = 1 hour left to play
11. The correct answer is B. $\frac{8}{15}$ can be broken down into 8 x $\frac{1}{15}$.
12. The correct answer is C. The base unit $\frac{1}{20}$ x 9 equals $\frac{9}{20}$.

13. The correct answer is D. The problem becomes $42 \times \frac{1}{13}$
14. The correct answer is A. He needs $\frac{3}{5} \times 10$ tubes of paint or $30 \times \frac{1}{5}$ tubes of paint. $\frac{30}{5} = 6$ tubes
15. The correct answer is C. The problem becomes $\frac{20}{100} + \frac{7}{100}$.
16. The correct answer is A. The problem becomes $\frac{80}{100} + \frac{15}{100}$.
17. The correct answer is D.
18. The correct answer is C.
19. The correct answer is B. $\frac{18}{100}$ is smaller than any other number.
20. The correct answer is A. $\frac{7}{10}$ is larger than $\frac{3}{10}$.

Chapter 4: Measurement & Data

4.1.A Solve problems involving measurement and conversion of measurements.

1. The correct answer is C. To find the answer, you multiply 60 by 40.
2. The correct answer is B. To find the answer, you divide 64 by 16 and remember, ounces convert to pounds.
3. The correct answer is A. To find the answer, you multiply 60 by 7 to get seconds.
4. The correct answer is D. To find the answer, you divide 8000 by 1000.
5. The correct answer is C. To find the answer, you multiply 3 by 1000.
6. The correct answer is C. To find the answer, you multiply 60 by 40.
7. The correct answer is B. To find the answer, you divide 8500 by 1000.
8. The correct answer is B. To find the answer, you divide 6700 by 1000.
9. The correct answer is A. To find the answer, you multiply 7.4 by 100.
10. The correct answer is D. To find the answer, you multiply 8 by 16.

ANSWER SHEET

4.1.B Solve problems involving measurement and conversion of measurements.

1. The correct answer is A. The easiest way to solve this problem is to convert the iced tea quarts to gallons.
2. The correct answer is A. Once you convert the iced tea to 12 quarts, and subtract 2, you get 12.
3. The correct answer is B. You add all the pounds and then add all the ounces and convert the extra ounces to pounds.
4. The correct answer is C. When you add all the minutes together, you get 150 minutes, which converts to 2 hours and 30 minutes.
5. The correct answer is D. First you must double their Monday distance and then you add that amount to the Monday and Friday amounts.
6. The correct answer is A. When you add all the items and complete the conversion, you get 12 lbs 4 oz.
7. The correct answer is B. The movies add up to 8 hours and 50 minutes, which you subtract from 10 hours.
8. The correct answer is A. You subtract 2 feet, 8 inches from 12 feet, 9 inches.
9. The correct answer is C. First you add 9 minutes to Christopher's time to find Emily's time and then you add that amount to Christopher's time to get the total.
10. The correct answer is B. You double 45 minutes then convert it to hours and minutes.

4.1.C Solve problems involving measurement and conversion of measurements.

1. The correct answer is D. First you double the length and subtract it from the perimeter. Then you divide that amount by 2 to get the width.
2. The correct answer is D. First you double the width and subtract it from the perimeter. Then you divide that amount by 2 to get the length.
3. The correct answer is A. You divide the area by the length to find the width.
4. The correct answer is B. You divide the area by the width to find the length.

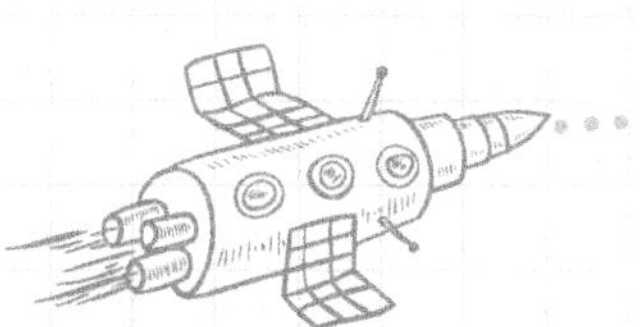

ANSWER SHEET

5. The correct answer is B. To find the perimeter, you take twice the width plus twice the length.
6. The correct answer is C. To find the area, you take the length times the width.
7. The correct answer is A. To find the width, you divide the area by the length.
8. The correct answer is A. First you double the width and subtract it from the perimeter. Then you divide that amount by 2 to get the length.
9. The correct answer is A. First you divide the area by the length to get the width. Then you double the length and the width and add them together to get the perimeter.
10. The correct answer is B. First you double the width and subtract it from the perimeter. Then you divide that amount by 2 to get the length. To find the area, you take the length times the width.

4.2 Represent and interpret data.

1. The correct answer is B because that makes where the lines divide the plot into 4s.

2. The correct answer is B because $\frac{4}{8}$ is the same as $\frac{1}{2}$.

3. The correct answer is D because the line is divided into 4s is the same as $\frac{1}{2}$.
4. The correct answer is A because the line is divided into 8s.
5. Correct answer: You should have placed you dot on the spot marked below.

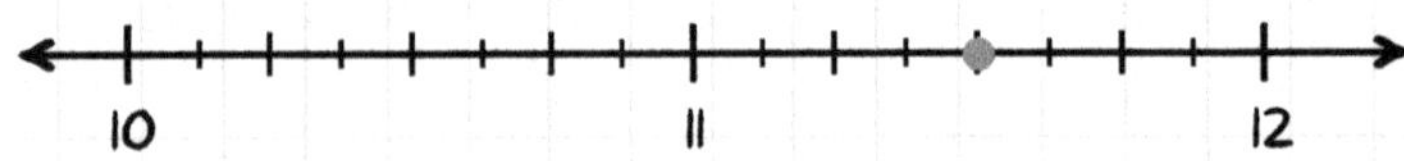

6. The correct answer is C because the value 15 is more than the range of the line plot.
7. The correct answer is D because the value 5 is less than the range of the line plot.
8. The correct answer is A because the value 35 is less than the range of the line plot.

ANSWER SHEET

9. The correct answer is B because a number more than 104 is off the line plot.
10. The correct answer is D because the value 68 is more than the range of the line plot.

4.3.A Understand concepts of angles and measure angles.

1. The correct answer is D because it is the only angle larger than 90°.
2. The correct answer is A because it is the angle closest to 0°.
3. The correct answer is C because it is the angle that measures 90° and is a corner.
4. The correct answer is B because the value is closest to 180 and it appears to be closest to a straight line.
5. 

4.3.B Understand concepts of angles and measure angles.

1. The correct answer is A because you name the angle by ray, center, ray.
2. The correct answer is B because you name the angle by ray, center, ray.
3. The correct answer is C because you name the angle by ray, center, ray.
4. The correct answer is B because you name the angle by ray, center, ray.
5. The correct answer is A because you name the angle by ray, center, ray.

4.3.C Understand concepts of angles and measure angles.

1. The correct answer is A because when you measure the angle, it is closest to 30°.
2. The correct answer is C because when you measure the angle, it is closest to 80°.
3. The correct answer is D because when you measure the angle, it is closest to 135°.
4. The correct answer should resemble an angle that looks like this:

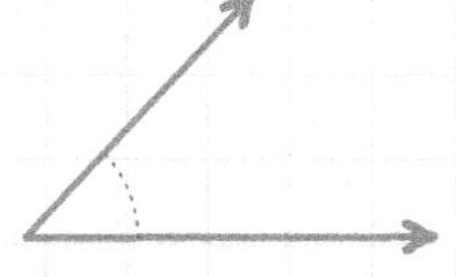

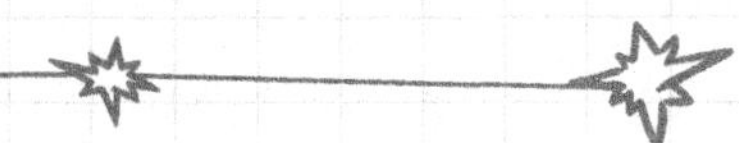

5. The correct answer should resemble an angle that looks like this:

4.3.D Understand concepts of angles and measure angles.

1. The correct answer is A because angle nop is the difference between the total angle and angle mon.
2. The correct answer is C because angle nop is the difference between the total angle and angle mon.
3. The correct answer is B because the answer is 110-90.
4. The correct answer is C because a 90° move on the clock will take the minute hand from 6 to 9.
5. The correct answer is C because a 180° move on the clock will take the minute hand from 3 to 9.

4.4. Chapter Test

1. The correct answer is D. To find the answer, you multiply 7.5 by 1000.
2. The correct answer is C. To find the answer, you multiply 60 by 40.
3. The correct answer is C. To find the answer, you multiply 632 by 0.01.
4. The correct answer is A. Once you convert all measurements to quarts and add them together, you get 30.
5. The correct answer is B. When you add six inches to 4 feet 8 inches, you get 5 feet 2 inches.
6. The correct answer is C. The difference between the weights is 11 ounces.
7. The correct answer is No because when you add up her hours, it adds up to over 10 hours.
8. The correct answer is D. When you subtract 20 feet, 8 inches from 28 feet, 6 inches, you get 7 feet, 10 inches.
9. The correct answer is A. First you double the length and subtract it from the perimeter. Then you divide that amount by 2 to get the width.

10. The correct answer is B. First you double the width and subtract it from the perimeter. Then you divide that amount by 2 to get the length. To get the area, you take the length times the width.
11. The correct answer is D because the value of $3\frac{5}{8}$ should be the closest point to 4.
12. The correct answer is C because the value 5 is less than the range of the line plot.
13. The correct answer is A because the value 35 is less than the range of the line plot.
14. The correct answer should resemble an angle that looks like this:

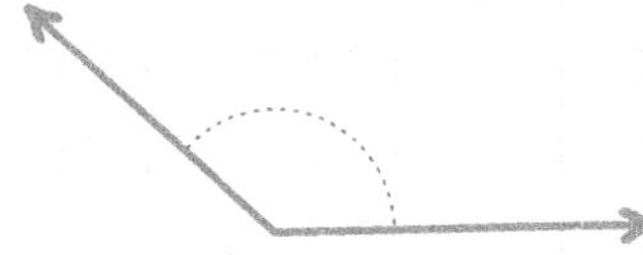

15. The correct answer is C because you name the angle by ray, center, ray.
16. The correct answer is B because you name the angle by ray, center, ray.
17. The correct answer is D because the angle measures closest to 85°.
18. The correct answer should resemble an angle that looks like this:

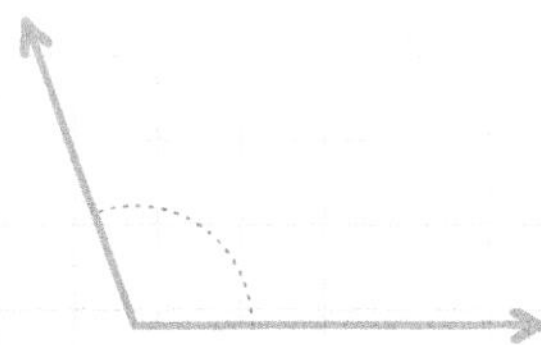

19. The correct answer is B because to find the answer you need to divide 100 by 2.
20. The correct answer is A because a 180° move on the clock will take the minute hand from 12 to 6.

Chapter 5: Geometry

5.1.A Draw and identify lines and angles, and classify shapes by properties of their lines and angles.

1. The correct answer is B. The other choices do not have two ends that continue forever.
2. The correct answer is D. The other choices do not have one end point and one end that continues forever.

3. The correct answer should look like this:

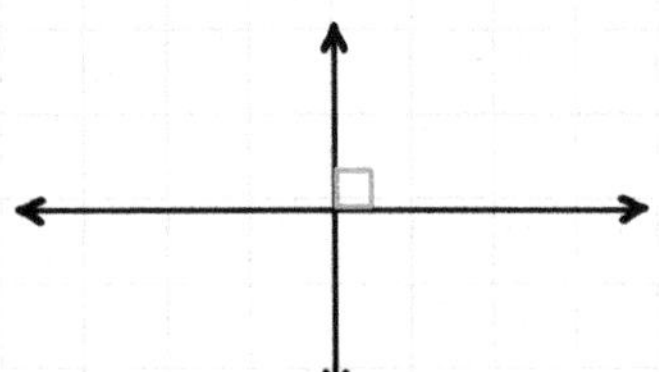

4. The correct answer is C because it is the only angle less than 90°.
5. The correct answer is A because it is the only angle that measures 90°.
6. The correct answer should look like this:

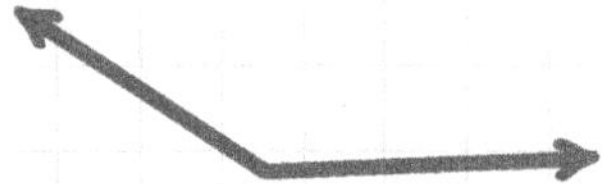

7. The correct answer is B. Points are expressed as one spot on a line.
8. Students should circle two lines which are across from each other that do not intersect.
9. Students should circle the two lines that make the triangles corner.
10. One possible shape students can draw is:

5.1.B Draw and identify lines and angles, and classify shapes by properties of their lines and angles.

1. The correct answer is B because a right triangle contains a 90°.
2. The correct answer is D. A hexagon contains at least 2 sets of parallel lines.
3. The correct answer is C. A rectangle contains both parallel and perpendicular lines.
4. The correct answer is B. A scalene triangle does not contain perpendicular or parallel lines.
5. The correct answer is A. A right triangle contains a 90° angle.
6. The correct answer is B. An obtuse triangle contains an angle above 90°.

7. The correct answer is D. A scalene triangle has three unequal sides.
8. The correct answer is C. The angle measurements of an equilateral triangle are always 60°.
9. The correct answer should look like this:

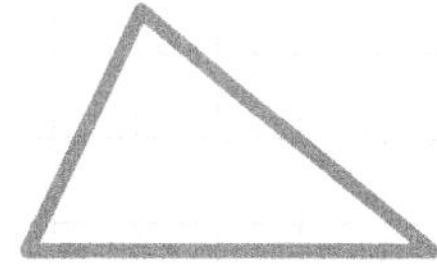

10. The correct answer should look like this:

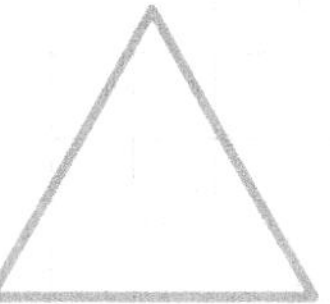

5.1.C Draw and identify lines and angles, and classify shapes by properties of their lines and angles.

1. The correct answer should look like this:

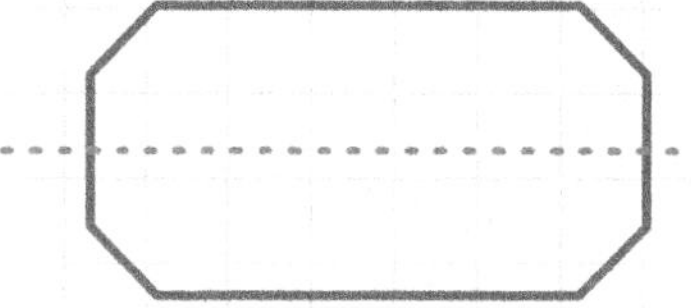

2. The correct answer should look like this:

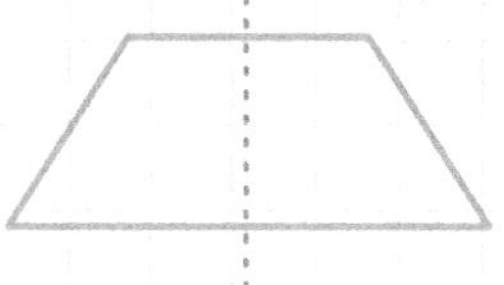

3. The correct answer is A. That is the only line that will divide the shape into two separate but equal shapes.
4. The correct answer is C. The line divides the triangle into two equal sides.
5. The correct answer is A. The line divides the arrow into two equal sides.

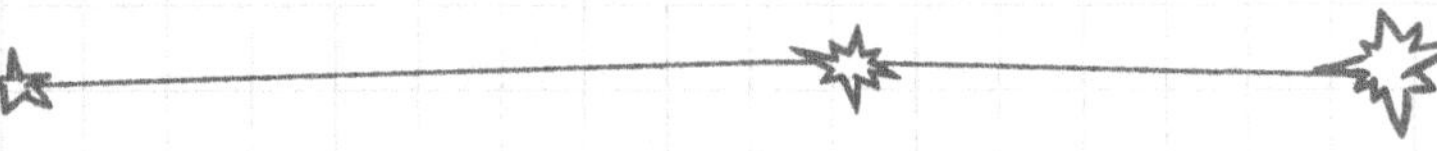

6. The correct answer should like this:

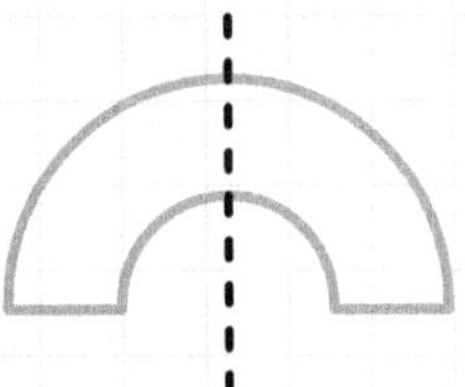

7. The correct answer is D. It is the only shape that can be divided into two separate but equal halves.
8. The correct answer is B. That shape cannot be divided into two or more equal parts.
9. The correct answer is a shape that can be divided into at least two equal parts.
10. The correct answer is a shape that cannot be divided into at least two equal parts.

5.2. Chapter Test

1. The correct answer is C. It is the only line that has an end on both sides.
2. The correct answer is A. The obtuse angle is over 90°.
3. Student responses should include a dot and a letter label.
4. The correct answer is C. It contains one set of parallel lines.
5. Student responses should not contain a right angle or lines that are equidistant and do not intersect.
6. The correct answer is C. This triangle is obtuse as it contains an angle greater than 90°.
7. The correct answer is A. This triangle contains two equal sides and angles.
8. The correct answer is A. An equilateral triangle contains three equal 60° angles.
9. The correct answer is B. It is the only triangle that contains an acute triangle.
10. Students responses should be a triangle that contains a 90° angle.
11. Students should draw a line where the shape will be divided into two or more equal parts.

12. The correct answer is D because it is the only shape that contains a line that divides it into two equal parts.
13. The correct answer is C because that is the only line that divides the shape into two equal parts.
14. The correct answer is A because that is the only shape that can be divided two equal parts.
15. The correct answer is B because that is the only shape that cannot be divided into two equal parts.

Chapter 6: Mixed Assessment

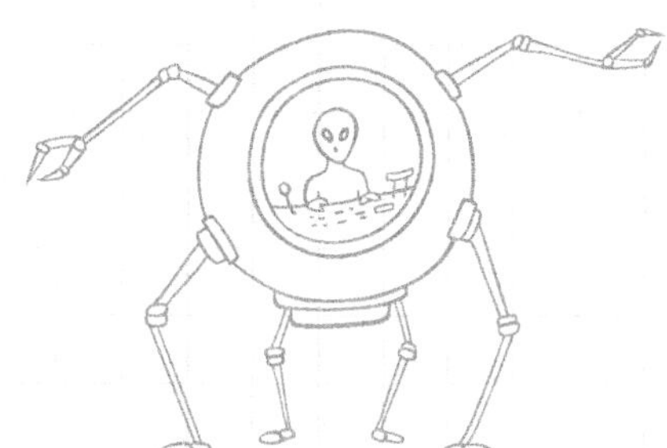

1. The correct answer is A. 8 x 6 =48
2. The correct answer is C. 8x4=32
3. The correct answer is B. 2 x 9 =18
4. The correct answer is D. 4 x 6 = 24
5. The correct answer is A. To find the correct number of trees, you need to take 10 x 3
6. The correct answer is C. 10 x 3 = 30
7. The correct answer is B. You need to take the total money divided by the cost per week.
8. The correct answer is B. 14÷2 =7
9. The correct answer is C. Both 1 and 8 can be multiplied by another number to get 64.
10. The correct answer is A. Both 5 and 8 can be multiplied by another number to get 40.
11. The correct answer is D. 3 is both a factor of 27 and a prime number.
12. The correct answer is D. 11 is both a factor of 88 and prime.
13. The correct answer is C. 10 is both a factor of 50 and composite.
14. The correct answer is B. 6 is both a factor of 42 and composite.
15. The correct answer is A. 5 x 1= 5, 5 x 2 = 10, 5 x 3 = 15
16. The correct answer is 17. Each term adds one more to the previous term.

ARGOPREP

ANSWER SHEET

17. The correct answer is -5, x 2. The first term is 5 more than the second term. The third term is double the second term and then the pattern is repeated.
18. The correct answer is a rectangle. The pattern is rectangle, triangle, circle, rectangle, circle, triangle.
19. The correct answer is 250. The pattern is -50, x 3. You take 300-50 to get 250 and that times 3 is 750.
20. The correct answer is -9, -8, -7. Each term is larger than the term before it by a certain amount and the pattern repeats.
21. The correct answer is B. 560,000 x 10 = 5,600,000.
22. The correct answer is D. 33,000 ÷ 10 = 3,300.
23. The correct answer is C. 5,671 is larger than 671, 5,067 and 567.
24. The correct answer is A. 4 million, 400,000 is smaller than 5 million, 40 million and 4 million, 500,000.
25. The correct answer is A. 10,681 is larger than 10,600; 10,081 and 1,681.
26. The correct answer is B. Four hundred is smaller than 700.
27. The correct answer is A. Ten thousand is larger than four thousand.
28. The correct answer is A. Eight hundred is larger than seven hundred.
29. The correct answer is D. Six stays the same because the tens place is 4.
30. The correct answer is C. Right rounds nineteen up to twenty.
31. The correct answer is C.
32. The correct answer is A.
33. The correct answer is A.
34. The correct answer is B.
35. The correct answer is B.
36. The correct answer is C.
37. The correct answer is 5,056
38. The correct answer is 53R3.
39. The correct answer is around 18 weeks. You should round 87 to 90 and then divide by 5.
40. The correct answer is around 37 dollars. You should divide 545 by 15 and then round up for a reasonable answer.

ANSWER SHEET

41. The correct answer is A. $4 \times 6 = 24$ and $7 \times 6 = 42$
42. The correct answer is D. While other answers are correct, if you use 10, you only need to find an equivalent fraction for one denominator.
43. The correct answer is C. They are equivalent fractions.
44. The correct answer is B. When adding 2 fractions, the denominators stay the same.
45. The correct answer is A. $\frac{13}{15}$ can be broken into $\frac{12}{15} + \frac{1}{15}$.
46. The correct answer is D. Your goal should be to have a 5 in the numerator.
47. The correct answer is B. Remember to combine both the mixed number and a fraction.
48. The correct answer is A. Remember, you need to borrow from the mixed number to subtract the fractions.
49. The correct answer is C. $\frac{1}{2} + \frac{1}{2} + \frac{1}{2} = 1\frac{1}{2} + 1 = 2\frac{1}{2}$ Then you need to subtract $6 - 2\frac{1}{2}$ to get the amount of time she has left to exercise.
50. The correct answer is A. $\frac{5}{7}$ can be broken down into $5 \times \frac{5}{7}$.
51. The correct answer is D. The base unit $\frac{1}{3} \times 8$ equals $\frac{8}{3}$.
52. The correct answer is B. To find how many eggs they need, you take the fraction times the whole.
53. The correct answer is C. The problem becomes $\frac{40}{100} + \frac{50}{100}$.
54. The correct answer is A.
55. The correct answer is B.
56. The correct answer is $\frac{16}{16} - \frac{5}{16} - \frac{3}{16}$ because you start with the whole and subtract the pie you served.
57. The correct answer is 0.27 . $\frac{27}{100}$ is the same as twenty-seven hundredths.

ARGOPREP

ANSWER SHEET

58. The correct answer is any number with **5** tenths or smaller.
59. The correct answer is >
60. The correct answer is $\frac{76}{100}$. Numbers to the hundredth power are the same as numbers with **100** in the denominator.

61. The correct answer is A. To find the answer, you divide **86** by **100**.
62. The correct answer is B. To find the answer, you multiply **8** by **60**.
63. The correct answer is A. To find the answer, you divide **5,612** by **1000**.
64. The correct answer is C. To find the width, you divide the area by the length.
65. The correct answer is A. To find the width, you take width times **2** and subtract it from the perimeter. Then you divide that amount by **2** for the length.
66. The correct answer is D. First you find the length by dividing the area by the width. Then you double the length and the width and add them together.
67. The correct answer is D. All the other choices are too large or too small for the line plot.
68. The correct answer is D. **56** is the largest whole number and **2/3** is the largest fraction.
69. The correct answer is B. **51** is the smallest whole number.
70. The correct answer is B. You name angles by ray, center, ray.
71. The correct answer is C. You name angles by ray, center, ray.
72. The correct answer is A. It is a straight line, which measures **180°**
73. The correct answer is B. **130 ÷ 2= 65**
74. The correct answer is A. Moving **90°** puts the minute hand on the **6**.
75. The correct answer is B. Moving **180°** puts the minute hand on the **6**.
76. The correct answer is A. Moving **360°** sends the minute hand all the way around the clock one time.

77\.

78\. The correct answer is 60°.

79\.

80\. The correct answer is 160°.

81\. The correct answer is A. The other choices do not have two ends that continue forever.

82\. The correct answer is D. The other choices do not have one end point and one end that continues forever.

83\. The correct answer is C. It is the only angle that is below 90°.

84\. The correct answer is B. It is the only angle that is above 90°.

85\. The correct answer is A. It is the only shape that have two lines that stay the same distance apart and do not intersect.

86\. The correct answer is D. All the other shapes contain 2 sets of lines that intersect at 87.

87\. The correct answer is B. It is formed by one line.

88\. The correct answer is C. An obtuse triangle contains one angle over 90°.

89\. The correct answer is D. A scalene triangle contains three different measurements.

90\. The correct answer is A. An equilateral triangle contains three angles that measure 60°.

91\. The correct answer is B. An right triangle contains an angle that measures 90°.

92\. The correct answer is C. An isosceles triangle has 2 equal angles.

93\. The correct answer is B. Angles in a triangle must add up to 180°

94\. The correct answer is A. The line divides the shape into two equal sides.

95\. The correct answer is D. That shape cannot be divided into two or more equal parts.

96\. Answer may vary.

97\. Answer may vary.

98\. Answer may vary.

99\. Answer may vary.

100\. Answer may vary.

Made in the USA
Coppell, TX
01 September 2021

61610635R00149